Selected Titles in This Series

(See the AMS catalog for earlier titles)

Basic Almost-Poised
Hypergeometric Series

MEMOIRS
of the
American Mathematical Society

Number 642

Basic Almost-Poised Hypergeometric Series

CHU Wenchang

September 1998 • Volume 135 • Number 642 (second of 5 numbers) • ISSN 0065-9266

American Mathematical Society
Providence, Rhode Island

1991 *Mathematics Subject Classification.*
Primary 33C20, 33D20; Secondary 05A19, 05A30.

Library of Congress Cataloging-in-Publication Data

Chu, Wenchang, 1958–
    Basic almost-poised hypergeometric series / Wenchang Chu.
        p. cm. — (Memoirs of the American Mathematical Society, ISSN 0065-9266 ; no. 642)
    "September 1998, volume 135, number 642 (second of 5 numbers)."
    Includes bibliographical references (p.   –   ).
    ISBN 0-8218-0811-7
    1. Hypergeometric series. I. Title. II. Series.
QA3.A57   no. 642
[QA295]
510 s–dc21
[515′.243]   98-25203
                                                                                      CIP

Memoirs of the American Mathematical Society

This journal is devoted entirely to research in pure and applied mathematics.

Subscription information. The 1998 subscription begins with volume 131 and consists of six mailings, each containing one or more numbers. Subscription prices for 1998 are $435 list, $348 institutional member. A late charge of 10% of the subscription price will be imposed on orders received from nonmembers after January 1 of the subscription year. Subscribers outside the United States and India must pay a postage surcharge of $30; subscribers in India must pay a postage surcharge of $43. Expedited delivery to destinations in North America $35; elsewhere $110. Each number may be ordered separately; *please specify number* when ordering an individual number. For prices and titles of recently released numbers, see the New Publications sections of the *Notices of the American Mathematical Society*.

Back number information. For back issues see the *AMS Catalog of Publications*.

Subscriptions and orders should be addressed to the American Mathematical Society, P. O. Box 5904, Boston, MA 02206-5904. *All orders must be accompanied by payment.* Other correspondence should be addressed to Box 6248, Providence, RI 02940-6248.

Memoirs of the American Mathematical Society is published bimonthly (each volume consisting usually of more than one number) by the American Mathematical Society at 201 Charles Street, Providence, RI 02904-2294. Periodicals postage paid at Providence, RI. Postmaster: Send address changes to Memoirs, American Mathematical Society, P. O. Box 6248, Providence, RI 02940-6248.

Contents

Abstract

A systematic treatment is presented for the evaluation of basic almost poised series. Around 200 identities are covered, among which most are believed to be new. Their connections with the q-Clausen formulae as well as Rogers-Ramanujan identities are sketched.

Key words and phrases: (basic) hypergeometric series, bilateral series, Chu-Vandermonde-Gauss formula, Saalschütz theorem, Whipple transform, Dougall-Dixon formulae, Jackson's q-analogue, Clausen product, Rogers-Ramanujan identities

Received by the editor March 8, 1996.

A mia moglie Colomba ed a mio figlio Yuang che, con il loro affettuoso sostegno, mi hanno dato la possibilità di dedicarmi completamente alla mia passione per la matematica.

0 Introduction

Following Bailey [1], the basic hypergeometric series reads as

$$
{}_{1+p}\phi_p\left[\begin{matrix} a_0, & a_1, & \cdots, & a_p \\ & b_1, & \cdots, & b_p \end{matrix} ; q, z\right] = \sum_{k=0}^{\infty} \frac{(a_0;q)_k\,(a_1;q)_k\,\cdots\,(a_p;q)_k}{(q;q)_k\,(b_1;q)_k\,\cdots\,(b_p;q)_k} z^k \tag{0.1}
$$

with the q-shifted factorial given by

$$
(x;q)_n = \frac{(x;q)_\infty}{(xq^n;q)_\infty} = \prod_{k=0}^{\infty}\frac{(1-xq^k)}{(1-xq^{k+n})} \tag{0.2a}
$$

$$
= \prod_{k=0}^{n-1}(1-xq^k), \quad (n = 1,2,\cdots) \tag{0.2b}
$$

where the first equality defines $(x;q)_n$ for all real n.

The bilateral basic hypergeometric series

$$
{}_r\psi_s\left[\begin{matrix} a_1, a_2, \cdots, a_r \\ b_1, b_2, \cdots, b_s \end{matrix} ; q, z\right] = \sum_{n=-\infty}^{+\infty}\{(-1)^n q^{\binom{n}{2}}\}^{s-r}\frac{(a_1;q)_k\,(a_2;q)_k\,\cdots\,(a_r;q)_k}{(b_1;q)_k\,(b_2;q)_k\,\cdots\,(b_s;q)_k} z^n \tag{0.3}
$$

corresponds to a bilateral summation (cf. [11]). To ensure convergence for the non-terminating series defined by (0.1) and (0.3), $|q| < 1$ will be assumed throughout the paper. When the "base" is q, we will omit to specify it explicitly in the basic hypergeometric notation of (0.1) and (0.3).

Almost poised series are defined, in view of Bressoud [4], by finite q-Dixon sums

$$
\omega_\delta\left[\begin{matrix} \alpha, & \gamma; & \varepsilon \\ b, & c; & n \end{matrix}\right] = {}_3\phi_2\left[\begin{matrix} q^{-\delta-2n}, & b, & c \\ & q^{\alpha-2n}/b, & q^{\gamma-2n}/c \end{matrix} ; \frac{q^{\varepsilon-n}}{bc}\right] \tag{0.4a}
$$

$$
\Omega_\delta\left[\begin{matrix} \alpha,\gamma,\beta,\rho; & \varepsilon \\ b, c, d, e; & n \end{matrix}\right] = {}_5\phi_4\left[\begin{matrix} q^{-\delta-2n}, & b, & c, & d, & e \\ & q^{\alpha-2n}/b, q^{\gamma-2n}/c, q^{\beta-2n}/d, q^{\rho-2n}/e \end{matrix} ; \frac{q^{\varepsilon-3n}}{bcde}\right] \tag{0.4b}
$$

where $\delta = 0$ or 1 corresponding to the finite Dixon sums of even and odd terms respectively. The parameters $\alpha, \gamma, \beta, \rho$ are integers confined from -5 to $+5$ so that the series concerned are "almost poised".

The compact notations for the q-factorial product and fraction

$$
[a, \ b, \ \cdots, \ c; \ q]_n = (a;q)_n\,(b;q)_n\cdots(c;q)_n \tag{0.5a}
$$

$$
\left[\begin{matrix} a, & b, \cdots, c \\ A, & B, \cdots, C \end{matrix} ; q\right]_n = \frac{(a;q)_n\,(b;q)_n\cdots(c;q)_n}{(A;q)_n\,(B;q)_n\cdots(C;q)_n} \tag{0.5b}
$$

will be used frequently in order to simplify the writing. Especially when $n = 1$, the base q and the subscript 1 in (0.5a) and (0.5b) will not be displayed.

1

The primary motivation for this work is due to the questions raised by Bressoud [4] in 1987. The partial result has been reported at the workshop on "Symbolic Computation in Combinatorics" at MSI (Cornell University, Ithaca, September, 1993). The author thanks D. M. Bressoud, Ira Gessel, S. Milne, P. Paule, V. Strehl and D. Zeilberger for their encouragement during the conference from whom the author has been stimulated to expand the work to the present version. Thanks should also go to Prof. Pier Vittorio Ceccherini, who has generously provided me the working facilities during the past few years, even we don't share much of the scientific interest. Without his support, my scientific career would have been broken down and I would never be able to conclude this work.

1 Basic Almost Poised Series

Recall the q-Saalschutz formula

$$
{}_3\phi_2\left[\begin{matrix} q^{-m}, & a, & b \\ & c, & q^{1-m}ab/c \end{matrix} ; q \right] = \frac{(c/a; q)_m\,(c/b; q)_m}{(c; q)_m\,(c/ab; q)_m}. \tag{1.1}
$$

For $\delta = 0, 1$ and integers ε, θ, α, γ, we make the following manipulation through interchanging the summation indices by introducing three additional integral parameters λ, μ, ν

$$
{}_5\phi_4\left[\begin{matrix} q^{-\delta-2n}, & w, & aq^\theta, & b, & c \\ & qw, & a, & q^{\alpha-2n}/b, & q^{\gamma-2n}/c \end{matrix} ; \frac{q^{\varepsilon-n}}{bc} \right] \tag{1.2a}
$$

$$
= \sum_{m=0}^{\delta+2n}\left[\begin{matrix} q^{-\delta-2n}, & w, & aq^\theta, & b, & c, & q^{\mu-2n}/b, & q^{\nu-2n}/c \\ q, & qw, & a, & bq^{\lambda-\mu}, & cq^{\lambda-\nu}, & q^{\alpha-2n}/b, & q^{\gamma-2n}/c \end{matrix} ; q \right]_m \tag{1.2b}
$$

$$
\times q^{m(\varepsilon+\lambda-\mu-\nu+n)}\,{}_3\phi_2\left[\begin{matrix} q^{-m}, & q^{\lambda-1+m-2n}, & q^{\mu+\nu-\lambda-2n}/bc \\ & q^{\mu-2n}/b, & q^{\nu-2n}/c \end{matrix} ; q \right] \tag{1.2c}
$$

$$
= \sum_{k=0}^{n+[\frac{1-\lambda}{2}]}\left[\begin{matrix} w, & aq^\theta, & b, & c, & q^{\mu+\nu-\lambda-2n}/bc \\ qw, & a, & bq^{\lambda-\mu}, & cq^{\lambda-\nu}, & q^{\alpha-2n}/b, & q^{\gamma-2n}/c \end{matrix} ; q \right]_k \tag{1.2d}
$$

$$
\times (-1)^k\, q^{k(\varepsilon+\lambda-\mu-\nu+n)-\binom{k}{2}}\,[q^{k+\lambda-1-2n}, q^{-\delta-2n}; q]_k/(q,; q)_k \tag{1.2e}
$$

$$
\times {}_8\phi_7\left[\begin{matrix} q^{\lambda-1-2n+2k}, & aq^{k+\theta}, & bq^k, & cq^k, & q^k w, \\ & aq^k, & bq^{\lambda-\mu+k}, & cq^{\lambda-\nu+k}, & q^{k+1}w, \end{matrix}\right. \tag{1.2f}
$$

$$
\left.\begin{matrix} q^{-\delta-2n+k}, & q^{\mu-2n+k}/b, & q^{\nu-2n+k}/c \\ q^{\lambda-1-2n+k}, & q^{\alpha-2n+k}/b, & q^{\gamma-2n+k}/c \end{matrix} ; q^{\varepsilon+\lambda-\mu-\nu+n-k} \right]^{(\delta+2n-k)} \tag{1.2g}
$$

where $\phi[\cdots]^{(n)}$ means the partial sum of the first $(n+1)$-terms of $\phi[\cdots]$-series.

Lemma (Gasper [10], see also Chu [9]) *For nonnegative integers* k_i *(* $i =$ *1, 2, $\cdots$, p), there is a partial-fraction formula*

$$
_{2+p}\phi_{1+p}\left[\begin{array}{c} q^{-n}, w, x_1 q^{k_1}, \cdots, x_p q^{k_p} \\ qw, \quad x_1, \quad \cdots, \quad x_p \end{array}; q^{1+n-m}\right] = w^m \left[\begin{array}{c} q \\ qw \end{array}; q\right]_n \prod_{i=1}^{p}\left[\begin{array}{c} x_i/w \\ x_i \end{array}; q\right]_{k_i} \quad (1.3)
$$

provided that $n \geq m \geq k = \sum_{i=1}^{p} k_i$.

PR-condition Parameter-restriction

$$
\lambda + \delta \leq 1 \tag{1.4a}
$$

$$
\theta \geq 0 \tag{1.4b}
$$

$$
\lambda \leq \mu \tag{1.4c}
$$

$$
\lambda \leq \nu \tag{1.4d}
$$

$$
\alpha \leq \mu \tag{1.4e}
$$

$$
\gamma \leq \nu. \tag{1.4f}
$$

In case **PR**-condition is satisfied, the $_8\phi_7$-series in the summation (1.2) may be evaluated, by the lemma, as a factorial product

$$
_8\phi_7\left[\begin{array}{cccccc} q^{\lambda-1-2n+2k}, & q^k w, & aq^{k+\theta}, & bq^k, & cq^k, & q^{-\delta-2n+k}, \\ & q^{k+1}w, & aq^k, & bq^{\lambda-\mu+k}, & cq^{\lambda-\nu+k}, & q^{\lambda-1-2n+k}, \end{array}\right. \tag{1.5a}
$$

$$
\left.\begin{array}{cc} q^{\mu-2n+k}/b, & q^{\nu-2n+k}/c \\ q^{\alpha-2n+k}/b, & q^{\gamma-2n+k}/c \end{array}; q^{\varepsilon+\lambda-\mu-\nu+n-k}\right]^{(1-\lambda+2n-2k)} \tag{1.5b}
$$

$$
= (wq^k)^{1+\lambda-\varepsilon+\delta+n}\left[\begin{array}{c} a/w \\ aq^k \end{array}; q\right]_\theta \left[\begin{array}{c} q^{\lambda-1-2n}/w \\ q^{\lambda-1-2n+k} \end{array}; q\right]_k \left[\begin{array}{c} q \\ wq^{k+1} \end{array}; q\right]_{\delta+2n-k} \tag{1.5c}
$$

$$
\times \left[\begin{array}{c} qw/b \\ q^{1-k}/b \end{array}; q\right]_{\mu-\lambda} \left[\begin{array}{c} qw/c \\ q^{1-k}/c \end{array}; q\right]_{\nu-\lambda} \left[\begin{array}{c} q^{\alpha-2n}/bw \\ q^{\alpha-2n+k}/b \end{array}; q\right]_{\mu-\alpha} \left[\begin{array}{c} q^{\gamma-2n}/cw \\ q^{\gamma-2n+k}/c \end{array}; q\right]_{\nu-\gamma} \tag{1.5d}
$$

provided $\theta - \delta - 2\lambda + 2\mu + 2\nu - \alpha - \gamma \leq 1 - \varepsilon - \lambda + \mu + \nu + n - k \leq 2n - 2k$.

Replacing the $_8\phi_7$-series in (1.2) by this evaluation and performing some modification, we get the main result of this section through which the evaluation of almost-poised series is reduced to computing a few terms of hypergeometric series.

Theorem *Denote by*

$$M = \theta - \delta + \mu + \nu - \lambda + \max\left\{\begin{array}{c} 1 - \varepsilon + \delta - \theta \\ -1 + \varepsilon - \alpha - \gamma \end{array}\right\}. \tag{1.6}$$

Choose λ, μ, ν so that the PR*-condition (1.4) is preserved. We have that*

$$_5\phi_4\left[\begin{array}{ccccc} q^{-\delta-2n}, & w, & aq^\theta, & b, & c \\ & qw, & a, & q^{\alpha-2n}/b, & q^{\gamma-2n}/c \end{array}; q^{\varepsilon-n}/bc\right] \tag{1.7a}$$

$$= \left[\begin{array}{c} a/w \\ a \end{array}; q\right]_\theta \left[\begin{array}{c} qw/b \\ q/b \end{array}; q\right]_{\mu-\lambda} \left[\begin{array}{c} qw/c \\ q/c \end{array}; q\right]_{\nu-\lambda} \left[\begin{array}{c} q^{\alpha-2n}/bw \\ q^{\alpha-2n}/b \end{array}; q\right]_{\mu-\alpha} \left[\begin{array}{c} q^{\gamma-2n}/cw \\ q^{\gamma-2n}/c \end{array}; q\right]_{\nu-\gamma} \tag{1.7b}$$

$$\times w^{1+\lambda-\varepsilon+\delta+n}\left[\begin{array}{c} q \\ qw \end{array}; q\right]_{\delta+2n} \sum_{k=0}^{n-M} q^k \left[\begin{array}{ccc} w, q^{\lambda-1-2n}/w, q^{\mu+\nu-\lambda-2n}/bc \\ q, q^{\mu-2n}/b, q^{\nu-2n}/c \end{array}; q\right]_k \tag{1.7c}$$

$$+ q^{n(\varepsilon+2\lambda-\mu-\nu-1)}\left[\begin{array}{c} q^{-\delta-2n}, q^{\mu+\nu-\lambda-2n}/bc \\ q^{\alpha-2n}/b, q^{\gamma-2n}/c \end{array}; q\right]_n \left[\begin{array}{ccccc} q^{2-\lambda}, w, aq^\theta, b, & c \\ q, qw, a, bq^{\lambda-\mu}, cq^{\lambda-\nu} \end{array}; q\right]_n \tag{1.7d}$$

$$\times \sum_{k=[\lambda/2]}^{M-1} \left[\begin{array}{ccccc} q^{2-\lambda+n}, q^{-n}/w, q^{1-n}/a, q^{1+\mu-\lambda-n}/b, q^{1+\nu-\lambda-n}/c \\ q^{1+\delta+n}, q^{1-n}/w, q^{1-\theta-n}/a, q^{1-n}/b, q^{1-n}/c \end{array}; q\right]_k \tag{1.7e}$$

$$\times (-1)^k q^{k(3-\varepsilon+\delta-\theta-\mu-\nu+\lambda+\alpha+\gamma)+\binom{k}{2}} \frac{[q^{-n}, bq^{1-\alpha+n}, cq^{1-\gamma+n}; q]_k}{(q^{2-\lambda}; q)_{2k} (bcq^{1+\lambda-\mu-\nu+n}; q)_k} \tag{1.7f}$$

$$\times {}_8\phi_7\left[\begin{array}{ccccccc} q^{\lambda-1-2k}, q^{n-k}w, aq^{\theta+n-k}, & bq^{n-k}, & cq^{n-k}, & q^{-\delta-n-k}, \\ q^{1+n-k}w, aq^{n-k}, bq^{\lambda-\mu+n-k}, cq^{\lambda-\nu+n-k}, q^{\lambda-1-n-k}, \end{array}\right. \tag{1.7g}$$

$$\left.\begin{array}{ccc} q^{\mu-n-k}/b, & q^{\nu-n-k}/c \\ q^{\alpha-n-k}/b, & q^{\gamma-n-k}/c \end{array}; q^{\varepsilon+\lambda-\mu-\nu+k}\right]^{(\delta+n+k)}. \tag{1.7h}$$

In this formula, the main part (1.7b-1.7c) of the series is the partial sum of a balanced series and the error-term (1.7d-1.7h) is derived under the replacement of $k \to n-k$ on the summation variable.

Choosing a, w, θ and λ, μ, ν so that the finite balanced series in (1.7c) is summable, we get the following specific summation-formulae which are especially useful in evaluating the almost-poised series with respect to various parameter-settings.

1.1   Main reduction formulae

In the theorem, take

$$\theta = 1 \tag{1.8a}$$

$$a = w \tag{1.8b}$$

which nullify the main part (1.7b-1.7c). The theorem reduces to

Proposition A *Assume* **M***-notation* (1.6) *and the* **PR***-condition* (1.4) *with* (1.4b) *replaced by* (1.8a). *We have*

$$\omega_\delta \begin{bmatrix} \alpha, & \gamma; & \varepsilon \\ b, & c; & n \end{bmatrix} = {}_3\phi_2 \begin{bmatrix} q^{-\delta-2n}, & b, & c & ; \dfrac{q^{\varepsilon-n}}{bc} \\ & q^{\alpha-2n}/b, q^{\gamma-2n}/c & \end{bmatrix} \tag{1.9a}$$

$$= q^{n(\varepsilon+2\lambda-\mu-\nu-1)} \begin{bmatrix} q^{2-\lambda}, & b, & c & ; q \\ q, & bq^{\lambda-\mu}, cq^{\lambda-\nu} & \end{bmatrix}_n \begin{bmatrix} q^{-\delta-2n}, & q^{\mu+\nu-\lambda-2n}/bc & ; q \\ q^{\alpha-2n}/b, & q^{\gamma-2n}/c & \end{bmatrix}_n \tag{1.9b}$$

$$\times \sum_{k=[\lambda/2]}^{M-1} (-1)^k \, q^{k(2-\varepsilon+\delta-\mu-\nu+\lambda+\alpha+\gamma)+\binom{k}{2}} / (q^{2-\lambda}; q)_{2k} \tag{1.9c}$$

$$\times \begin{bmatrix} q^{-n}, & q^{2-\lambda+n}, & q^{1+\mu-\lambda-n}/b, & q^{1+\nu-\lambda-n}/c, & bq^{1-\alpha+n}, & cq^{1-\gamma+n} \\ q^{1+\delta+n}, & q^{1-n}/b, & q^{1-n}/c, & bcq^{1+\lambda-\mu-\nu+n} & ; q \end{bmatrix}_k \tag{1.9d}$$

$$\times {}_6\phi_5 \begin{bmatrix} q^{\lambda-1-2k}, & bq^{n-k}, & cq^{n-k}, & q^{-\delta-n-k}, \\ & bq^{\lambda-\mu+n-k}, & cq^{\lambda-\nu+n-k}, & q^{\lambda-1-n-k}, \end{bmatrix} \tag{1.9e}$$

$$\begin{bmatrix} q^{\mu-n-k}/b, & q^{\nu-n-k}/c & ; q^{\varepsilon+\lambda-\mu-\nu+k} \\ q^{\alpha-n-k}/b, & q^{\gamma-n-k}/c & \end{bmatrix}^{(1-\lambda+2k)} \tag{1.9f}$$

In the theorem, take

$$\theta = 1 \tag{1.10a}$$

$$\lambda + \delta = 2 \tag{1.10b}$$

$$a = w = q^{-\delta-2n} \tag{1.10c}$$

which cancel two pairs of numerator-denominator parameters. The main part (1.7b-1.7c) in the formula becomes

$$q^{(2n+\delta)\left(\frac{3}{2}(1-\delta)+\varepsilon-\alpha-\gamma\right)} \begin{bmatrix} b, & c, & bq^{1-\delta-\mu}, cq^{1-\delta-\nu} & ; q \\ bq^{1-\delta-\alpha}, cq^{1-\delta-\gamma}, bq^{2-\delta-\mu}, cq^{2-\delta-\nu} & \end{bmatrix}_{\delta+2n}$$

$$\times (-1)^{1+\delta} \left\{ \sum_{k=0}^{n-M} - \sum_{k=0}^{\delta+2n} \right\} \begin{bmatrix} q^{-\delta-2n}, & q^{\delta+\mu+\nu-2-2n}/bc & ; q \\ q^{\mu-2n}/b, & q^{\nu-2n}/c & \end{bmatrix}_k q^k$$

where the second sum consists of the singular terms missed from the replacement of ${}_8\phi_7[\cdots]^{(\delta+2n-k)}$ in (1.2) by the ${}_8\phi_7[\cdots]^{(1-\lambda+2n-2k)}$-evaluation in (1.5) because (1.4a) is not consistent with (1.10b). The reversal of the summation can be evaluated by the Saalschütz theorem (1.1) which leads us to

Proposition B *Assume* **M**-*notation* (1.6) *and the* **PR**-*condition* (1.4) *with* (1.4a) *and* (1.4b) *replaced by* (1.10b) *and* (1.10a), *respectively. We have*

$$\omega_\delta \begin{bmatrix} \alpha, & \gamma; & \varepsilon \\ b, & c; & n \end{bmatrix} = {}_3\phi_2 \begin{bmatrix} q^{-\delta-2n}, & b, & c & \dfrac{q^{\varepsilon-n}}{bc} \\ & q^{\alpha-2n}/b, & q^{\gamma-2n}/c & ; \end{bmatrix} \tag{1.11a}$$

$$= (-1)^\delta q^{\binom{2n+\delta}{}(\frac{1-\delta}{2}+\varepsilon-\delta-\alpha-\gamma)} \begin{bmatrix} b, & c \\ bq^{1-\delta-\alpha}, cq^{1-\delta-\gamma}; q \end{bmatrix}_{\delta+2n} \begin{bmatrix} q^{-\delta-2n}, q^{\delta+\mu+\nu-2-2n}/bc \\ q^{\mu-1-2n}/b, \ q^{\nu-1-2n}/c \end{bmatrix}_{n-M+1} \tag{1.11b}$$

$$+ \, q^{n\,(3-2\delta+\varepsilon-\mu-\nu)} \begin{bmatrix} q^\delta, & b, & c \\ q, bq^{2-\delta-\mu}, cq^{2-\delta-\nu}; q \end{bmatrix}_n \begin{bmatrix} q^{-\delta-2n}, \ q^{\delta+\mu+\nu-2-2n}/bc \\ q^{\alpha-2n}/b, \ q^{\gamma-2n}/c \end{bmatrix}_n \tag{1.11c}$$

$$\times \sum_{k=1-\delta}^{M-1} (-1)^k q^{k(4-\varepsilon-\mu-\nu+\alpha+\gamma)+\binom{k}{2}} / \, (q^\delta; q)_{2k} \tag{1.11d}$$

$$\times \begin{bmatrix} q^{-n}, q^{\delta+n}, & q^{\delta+\mu-1-n}/b, & q^{\delta+\nu-1-n}/c, & bq^{1-\alpha+n}, & cq^{1-\gamma+n} \\ q^{1+\delta+n}, & q^{1-n}/b, & q^{1-n}/c, & bcq^{3-\delta-\mu-\nu+n} \end{bmatrix}_k \tag{1.11e}$$

$$\times {}_6\phi_5 \begin{bmatrix} q^{1-\delta-2k}, & bq^{n-k}, & cq^{n-k}, & q^{-\delta-n-k}, \\ & bq^{2-\delta-\mu+n-k}, & cq^{2-\delta-\nu+n-k}, & q^{1-\delta-n-k}, \end{bmatrix} \tag{1.11f}$$

$$\begin{matrix} q^{\mu-n-k}/b, & q^{\nu-n-k}/c \\ q^{\alpha-n-k}/b, & q^{\gamma-n-k}/c \end{matrix} \; ; \, q^{2+\varepsilon-\delta-\mu-\nu+k} \end{bmatrix}^{(\delta-1+2k)} \tag{1.11g}$$

In the theorem, take

$$\theta = 1 \tag{1.12a}$$

$$1 + \mu = \lambda \tag{1.12b}$$

$$a = w = b \tag{1.12c}$$

which cancel two pairs of numerator-denominator parameters and again replace $_8\phi_7[\cdots]^{(\delta+2n-k)}$ in (1.2) by the $_8\phi_7[\cdots]^{(1-\lambda+2n-2k)}$-evaluation in (1.5). The main part (1.7b-1.7c) in the formula becomes

$$\begin{bmatrix} qb/c \\ q/c \end{bmatrix}_{\nu-\lambda} \begin{bmatrix} q^{\alpha-2n}/b^2 \\ q^{\alpha-2n}/b \end{bmatrix}_{\lambda-1-\alpha} \begin{bmatrix} q^{\gamma-2n}/bc \\ q^{\gamma-2n}/c \end{bmatrix}_{\nu-\gamma}$$

$$\times b^{\lambda-\varepsilon+\delta+n} \begin{bmatrix} q \\ bq \end{bmatrix}_{\delta+2n} \sum_{k=0}^{n-M} \begin{bmatrix} b, & q^{\nu-1-2n}/bc \\ q, & q^{\nu-2n}/c \end{bmatrix}_k q^k \, .$$

Evaluating the sum by the Saalschütz theorem (1.1) leads us to

Proposition C *Assume* **M**-*notation* (1.6) *and the* **PR**-*condition* (1.4) *with* (1.4b) *and* (1.4c) *repaced by* (1.12a) *and* (1.12b), *respectively. We have*

$$\omega_\delta \begin{bmatrix} \alpha, & \gamma; & \varepsilon \\ b, & c; & n \end{bmatrix} = {}_3\phi_2 \begin{bmatrix} q^{-\delta-2n}, & b, & c \\ & q^{\alpha-2n}/b, & q^{\gamma-2n}/c \end{bmatrix} ; \frac{q^{\varepsilon-n}}{bc} \end{bmatrix} \tag{1.13a}$$

$$= b^{\lambda-\varepsilon-M} \begin{bmatrix} qb/c \\ q/c \end{bmatrix} ; q \end{bmatrix}_{\nu-\lambda} \begin{bmatrix} q^{\alpha-2n}/b^2 \\ q^{\alpha-2n} \end{bmatrix} ; q \end{bmatrix}_{\lambda-1-\alpha} \tag{1.13b}$$

$$\times \begin{bmatrix} q^{-\delta-2n} \\ q^{-\delta-2n}/b \end{bmatrix} ; q \end{bmatrix}_{\delta+n+M} \begin{bmatrix} q^{\gamma-2n}/bc \\ q^{\gamma-2n}/c \end{bmatrix} ; q \end{bmatrix}_{\nu-\gamma+n-M} \tag{1.13c}$$

$$+ q^{n(\varepsilon+\lambda-\nu)} \begin{bmatrix} q^{2-\lambda}, b, & c \\ q, & bq, cq^{\lambda-\nu} \end{bmatrix} ; q \end{bmatrix}_n \begin{bmatrix} q^{-\delta-2n}, q^{\nu-\lambda-1-2n}/bc \\ q^{\alpha-2n}/b, q^{\gamma-2n}/c \end{bmatrix} ; q \end{bmatrix}_n \tag{1.13d}$$

$$\times \sum_{k=[\lambda/2]}^{M-1} (-1)^k q^{k(3-\varepsilon+\delta-\nu+\alpha+\gamma)+\binom{k}{2}} / (q^{2-\lambda}; q)_{2k} \tag{1.13e}$$

$$\times \begin{bmatrix} q^{-n}, & q^{2-\lambda+n}, & q^{-n}/b, q^{1+\nu-\lambda-n}/c, & bq^{1-\alpha+n}, & cq^{1-\gamma+n} \\ q^{1+\delta+n}, & q^{1-n}/b, & q^{1-n}/c, & bcq^{2-\nu+n} \end{bmatrix} ; q \end{bmatrix}_k \tag{1.13f}$$

$$\times {}_6\phi_5 \begin{bmatrix} q^{\lambda-1-2k}, & bq^{n-k}, & cq^{n-k}, & q^{-\delta-n-k}, \\ & bq^{1+n-k}, & cq^{\lambda-\nu+n-k}, & q^{\lambda-1-n-k}, \end{bmatrix} \tag{1.13g}$$

$$\begin{matrix} q^{\lambda-1-n-k}/b, & q^{\nu-n-k}/c \\ q^{\alpha-n-k}/b, & q^{\gamma-n-k}/c \end{matrix} ; q^{1+\varepsilon-\nu+k} \end{bmatrix}^{(1-\lambda+2k)} \tag{1.13h}$$

In the theorem, take

$$\theta = 1 \tag{1.14a}$$
$$1 + \mu = \alpha \tag{1.14b}$$
$$a = w = q^{\alpha-1-2n}/b \tag{1.14c}$$

which cancel two pairs of numerator-denominator parameters and again replace ${}_8\phi_7[\cdots]^{(\delta+2n-k)}$ in (1.2) by the ${}_8\phi_7[\cdots]^{(1-\lambda+2n-2k)}$-evaluation in (1.5). The main part (1.7b-1.7c) of the formula becomes

$$\begin{bmatrix} \frac{q}{q^{\alpha-2n}/b}; q \end{bmatrix}_{\delta+2n} \begin{bmatrix} q^{1+\gamma-\alpha}b/c \\ q^{\gamma-2n}/c \end{bmatrix} ; q \end{bmatrix}_{\nu-\gamma} \begin{bmatrix} q^{\alpha-2n}/b^2 \\ q/b \end{bmatrix} ; q \end{bmatrix}_{\alpha-\lambda-1} \begin{bmatrix} q^{\alpha-2n}/bc \\ q/c \end{bmatrix} ; q \end{bmatrix}_{\nu-\lambda}$$

$$\times (bq^{1-\alpha+2n})^{\varepsilon-\delta-\lambda-1-n} \sum_{k=0}^{n-M} \begin{bmatrix} bq^{\lambda-\alpha}, q^{\alpha+\nu-\lambda-1-2n}/bc \\ q, & q^{\nu-2n}/c \end{bmatrix} ; q \end{bmatrix}_k q^k .$$

Evaluating the sum by the Saalschütz theorem (1.1) leads us to

Proposition D *Assume **M**-notation (1.6) and the **PR**-condition (1.4) with (1.4b)*
and (1.4e) replaced by (1.14a) and (1.14b), respectively. We have

$$\omega_\delta \begin{bmatrix} \alpha, & \gamma; & \varepsilon \\ b, & c; & n \end{bmatrix} = {}_3\phi_2 \begin{bmatrix} q^{-\delta-2n}, & b, & c & q^{\varepsilon-n} \\ & q^{\alpha-2n}/b, & q^{\gamma-2n}/c & ; \frac{q^{\varepsilon-n}}{bc} \end{bmatrix} \tag{1.15a}$$

$$= (-1)^\delta \, q^{\lambda(n-M)+(\delta+M)(1+2n)} \frac{(q^{1+\gamma-\alpha}b/c; q)_{\nu-\gamma} \, (q^{\alpha-2n}/b^2; q)_{\alpha-\lambda-1}}{(q/b; q)_{\alpha+\delta-1} \, (q/c; q)_{\nu-\lambda}} \tag{1.15b}$$

$$\times (bq^{1-\alpha+2n})^{\varepsilon-\delta-\lambda-1-M} \frac{(q^{-\delta-2n}; q)_{\delta+n+M} \, (q^{\alpha-2n}/bc; q)_{\nu-\lambda+n-M}}{(q^{\alpha-2n}/b; q)_{-\lambda+n+M} \, (q^{\gamma-2n}/c; q)_{\nu-\gamma+n-M}} \tag{1.15c}$$

$$+ \, q^{n(\varepsilon+2\lambda-\alpha-\nu)} \begin{bmatrix} q^{2-\lambda}, & b, & c \\ q, & bq^{1+\lambda-\alpha}, & cq^{\lambda-\nu} \end{bmatrix}_n \begin{bmatrix} q^{-\delta-2n}, & q^{\alpha+\nu-\lambda-1-2n}/bc \\ q^{\alpha-2n}/b, & q^{\gamma-2n}/c \end{bmatrix}_n \tag{1.15d}$$

$$\times \sum_{k=[\lambda/2]}^{M-1} (-1)^k q^{k(3-\varepsilon+\delta+\lambda-\nu+\gamma)+\binom{k}{2}} / \, (q^{2-\lambda}; q)_{2k} \tag{1.15e}$$

$$\times \begin{bmatrix} q^{-n}, & q^{2-\lambda+n}, & q^{\alpha-\lambda-n}/b, & q^{1+\nu-\lambda-n}/c, & bq^{1-\alpha+n}, & cq^{1-\gamma+n} \\ q^{1+\delta+n}, & q^{1-n}/b, & q^{1-n}/c, & bcq^{2+\lambda-\alpha-\nu+n} \end{bmatrix}_k \tag{1.15f}$$

$$\times {}_6\phi_5 \begin{bmatrix} q^{\lambda-1-2k}, & bq^{n-k}, & cq^{n-k}, & q^{-\delta-n-k}, \\ & bq^{1+\lambda-\alpha+n-k}, & cq^{\lambda-\nu+n-k}, & q^{\lambda-1-n-k}, \end{bmatrix} \tag{1.15g}$$

$$\begin{matrix} q^{\alpha-1-n-k}/b, & q^{\nu-n-k}/c \\ q^{\alpha-n-k}/b, & q^{\gamma-n-k}/c \end{matrix} ; \, q^{1+\varepsilon+\lambda-\alpha-\nu+k} \end{bmatrix}^{(1-\lambda+2k)} \tag{1.15h}$$

By means of Proposition **A-D**, we may derive closed formulas for almost poised
series with various specific parameters. The complete list obtained by the author
up to now is displayed in the next two sections, where the propositions used and
the references related are indicated under the *Note* for each formula.

In order to simplify writing, we will use the following notation.

Definition 1 *Extra factor of the almost-poised series*

$$\chi_\delta \begin{bmatrix} \alpha, & \gamma; & \varepsilon \\ b, & c; & n \end{bmatrix} = \omega_\delta \begin{bmatrix} \alpha, & \gamma; & \varepsilon \\ b, & c; & n \end{bmatrix} / \begin{bmatrix} q^{-\delta-2n}, & q^{[\alpha,\gamma]-2n}/bc \\ q^{\alpha-2n}/b, & q^{\gamma-2n}/c \end{bmatrix}_n \tag{1.16}$$

where $[\alpha, \gamma] = (\alpha + \gamma + |\alpha - \gamma|)/2.$

1.2 Basic almost poised formulas when $\delta = 0$

α	γ	ε	$\chi_0[\alpha, \gamma; \varepsilon]$	note
1	1	0	$q^{-n} + q^{-2n}\begin{bmatrix} q^n,\, bq^n,\, cq^n \\ bcq^n \end{bmatrix}$	$A: \lambda = 1$ $\mu = 1$ $\nu = 1$
		1	1	
		2	q^n	
		3	$q^{2n} + q^n \begin{bmatrix} q^n,\, bq^n,\, cq^n \\ bcq^n \end{bmatrix}$	(cf. [2, 4, 5] & [14, 16])
0	2	1	$1 + q^{-n}\begin{bmatrix} q^n,\, b \\ c/q,\, bcq^{n-1} \end{bmatrix}$	$A: \lambda = 1$ $\mu = 1$ $\nu = 2$
		2	$1 + c^2 q^{n-2}\begin{bmatrix} q^n,\, b \\ c/q,\, bcq^{n-1} \end{bmatrix}$	(cf. [4])
1	2	1	$q^{-2n}\begin{bmatrix} cq^{n-1} \\ c/q \end{bmatrix} \times \left\{ 1 - bq^n \begin{bmatrix} q^n,\, c/q \\ bcq^{n-1} \end{bmatrix} \right\}$	$A: \lambda = 1$ $\mu = 1$ $\nu = 2$
		2	$\begin{bmatrix} cq^{n-1} \\ c/q \end{bmatrix}$	$B: \lambda = 2$ $\mu = 2$ $\nu = 2$
		3	$q^{2n}\begin{bmatrix} cq^{n-1} \\ c/q \end{bmatrix} \times \left\{ 1 + q^{-n}\begin{bmatrix} q^n,\, c/q \\ bcq^{n-1} \end{bmatrix} \right\}$	(cf. [15])
0	1	0	$1 + q^{-n}\begin{bmatrix} q^n,\, b \\ bcq^n \end{bmatrix}$	$A: \lambda = 1$ $\mu = 1$ $\nu = 1$
		1	1	
		2	$1 - cq^n \begin{bmatrix} q^n,\, b \\ bcq^n \end{bmatrix}$	(cf. [4, 15])
0	3	2	$1 + q^{-n}\begin{bmatrix} q^n,\, b,\, c^3 q^{2n-5} \\ c/q,\, c/q^2,\, bcq^{n-2} \end{bmatrix}$	$A: \lambda = 1$ $\mu = 1$ $\nu = 3$
2	-1	1	$q^{-n}\begin{bmatrix} bq^{n-1} \\ b/q \end{bmatrix} \times \left\{ 1 - b/q \begin{bmatrix} q^n,\, q^{n+2}c^2/b \\ cq^{n+1},\, bcq^{n-1} \end{bmatrix} \right\}$	$A: \lambda = 1$ $\mu = 2$ $\mu = 1$

α	γ	ε	$\chi_0[\alpha,\gamma;\varepsilon]$	note
2	2	1	$q^{-3n}\begin{bmatrix}bq^{n-1},\,cq^{n-1}\\b/q,\,c/q\end{bmatrix}\times\left\{1+q^{n+1}\begin{bmatrix}q^n,\,b/q,\,c/q\\bcq^{n-1}\end{bmatrix}\right\}$	$A:\lambda=1$ $\mu=2$ $\nu=2$
		2	$q^{-n}\begin{bmatrix}bq^{n-1},\,cq^{n-1}\\b/q,\,c/q\end{bmatrix}$	
		3	$q^{n}\begin{bmatrix}bq^{n-1},\,cq^{n-1}\\b/q,\,c/q\end{bmatrix}$	$B:\lambda=2$ $\mu=2$ $\nu=2$
		4	$q^{3n}\begin{bmatrix}bq^{n-1},\,cq^{n-1}\\b/q\ c/q\end{bmatrix}\times\left\{1+q^{-n}\begin{bmatrix}q^n,\,b/q,\,c/q\\bcq^{n-1}\end{bmatrix}\right\}$	(cf. [4])
0	0	-1	$1+q^{-n}\begin{bmatrix}q^n,\,b,\,c\\bcq^{n+1}\end{bmatrix}$	$A:\lambda=0$ $\mu=0$ $\nu=0$
		0	1	
		1	1	
		2	$1+q^{n+1}\begin{bmatrix}q^n,\,b,\,c\\bcq^{n+1}\end{bmatrix}$	(cf. [4])
1	3	2	$q^{-n}\begin{bmatrix}cq^{n-1},\,cq^{n-2}\\c/q,\,c/q^2\end{bmatrix}\times\left\{1+q^{-n}\begin{bmatrix}q^n,\,bq^n\\cq^{n-1},\,bcq^{n-2}\end{bmatrix}\right\}$	$A:\lambda=1$ $\mu=1$ $\nu=3$
		3	$\begin{bmatrix}cq^{n-1},\,cq^{n-2}\\c/q,\,c/q^2\end{bmatrix}\times\left\{1+c^2q^{n-3}\begin{bmatrix}q^n,\,bq^n\\cq^{n-1},\,bcq^{n-2}\end{bmatrix}\right\}$	$B:\lambda=2$ $\mu=2$ $\nu=3$
1	-1	0	$q^{-n}+c^2q\begin{bmatrix}q^n,\,bq^n\\cq^{n+1},\,bcq^n\end{bmatrix}$	$A:\lambda=1$ $\mu=1$ $\nu=1$
		1	$1+q^{-n}\begin{bmatrix}q^n,\,bq^n\\cq^{n+1},\,bcq^n\end{bmatrix}$	

α	γ	ε	$\chi_0[\alpha,\gamma;\varepsilon]$	note
2	3	2	$q^{-3n}\begin{bmatrix} bq^{n-1},\ cq^{n-2},\ cq^{2n-1} \\ b/q,\ c/q,\ c/q^2 \end{bmatrix} \times \left\{ 1 + q^{n+1}\begin{bmatrix} q^n,\ b/q,\ c/q^2 \\ cq^{2n-1},\ bcq^{n-2} \end{bmatrix}\right\}$	$A:\lambda=1$ $\mu=2$ $\nu=3$
		3	$q^{-n}\begin{bmatrix} bq^{n-1},\ cq^{n-2},\ cq^{2n-1} \\ b/q,\ c/q,\ c/q^2 \end{bmatrix}$	
		4	$q^{n}\begin{bmatrix} bq^{n-1},\ cq^{n-2},\ cq^{2n-1} \\ b/q,\ c/q,\ c/q^2 \end{bmatrix} \times \left\{ 1 - cq^{n-1}\begin{bmatrix} q^n,\ b/q,\ c/q^2 \\ cq^{2n-1},\ bcq^{n-2} \end{bmatrix}\right\}$	$B:\lambda=2$ $\mu=2$ $\nu=3$
0	-1	-1	$q^{-n}\begin{bmatrix} cq^{2n+1} \\ cq^{n+1} \end{bmatrix} \times \left\{ 1 - cq^{n+1}\begin{bmatrix} q^n,\ b,\ c \\ cq^{2n+1},\ bcq^{n+1} \end{bmatrix}\right\}$	$A:\lambda=0$ $\mu=0$ $\nu=0$
		0	$q^{-n}\begin{bmatrix} cq^{2n+1} \\ cq^{n+1} \end{bmatrix}$	
		1	$q^{-n}\begin{bmatrix} cq^{2n+1} \\ cq^{n+1} \end{bmatrix} \times \left\{ 1 + q^{n+1}\begin{bmatrix} q^n,\ b,\ c \\ cq^{2n+1},\ bcq^{n+1} \end{bmatrix}\right\}$	
1	4	3	$q^{-2n}\begin{bmatrix} cq^{2n-1},\ cq^{n-2},\ cq^{n-3},\ bcq^{2n-3} \\ c/q,\ c/q^2,\ c/q^3,\ bcq^{n-3} \end{bmatrix} \times \left\{ 1 - bq^{n}\begin{bmatrix} q^n,\ -cq^{n-2},\ q^{-2}c/b \\ cq^{2n-1},\ bcq^{2n-3} \end{bmatrix}\right\}$	$A:\lambda=1$ $\mu=1$ $\nu=4$
1	-2	0	$q^{-2n}\begin{bmatrix} cq^{2n+2},\ bcq^{2n} \\ cq^{n+2},\ bcq^{n} \end{bmatrix} \times \left\{ 1 - bq^{n}\begin{bmatrix} q^n,\ -cq^{n+1},\ qc/b \\ cq^{2n+2},\ bcq^{2n} \end{bmatrix}\right\}$	$A:\lambda=1$ $\mu=1$ $\nu=1$

α	γ	ε	$\chi_0[\alpha, \gamma; \varepsilon]$	note
3	3	3	$q^{-2n} \begin{bmatrix} bq^{n-1},\ bq^{n-2},\ cq^{n-1},\ cq^{n-2} \\ b/q,\ b/q^2,\ c/q,\ c/q^2 \end{bmatrix}$ $\times \left\{ 1 + q^{n+1} \begin{bmatrix} q^n,\ bcq^{n-3} \\ bq^{n-1},\ cq^{n-1} \end{bmatrix} \right\}$	$A: \lambda = 1$ $\mu = 3$ $\nu = 3$
		4	$q^n \begin{bmatrix} bq^{n-1},\ bq^{n-2},\ cq^{n-1},\ cq^{n-2} \\ b/q,\ b/q^2,\ c/q,\ c/q^2 \end{bmatrix}$ $\times \left\{ 1 + q^{-n} \begin{bmatrix} q^n,\ bcq^{n-3} \\ bq^{n-1},\ cq^{n-1} \end{bmatrix} \right\}$	$B: \lambda = 3$ $\mu = 3$ $\nu = 3$
-1	-1	0	$q^{-n} + q \begin{bmatrix} q^n,\ bcq^{n+1} \\ bq^{n+1},\ cq^{n+1} \end{bmatrix}$	$A: \lambda = -1$ $\mu = -1$ $\nu = -1$
		-1	$1 + q^{-n} \begin{bmatrix} q^n,\ bcq^{n+1} \\ bq^{n+1},\ cq^{n+1} \end{bmatrix}$	
2	4	3	$q^{-3n} \begin{bmatrix} bq^{n-1},\ cq^{2n-1},\ cq^{n-2},\ cq^{n-3},\ bcq^{2n-3} \\ b/q,\ c/q,\ c/q^2,\ c/q^3,\ bcq^{n-3} \end{bmatrix}$ $\times \left\{ 1 + q^{n+1} \begin{bmatrix} q^n,\ b/q,\ -cq^{n-2} \\ cq^{2n-1},\ bcq^{2n-3} \end{bmatrix} \right\}$	$A: \lambda = 1$ $\mu = 2$ $\nu = 4$
		4	$q^{-n} \begin{bmatrix} bq^{n-1},\ cq^{2n-1},\ cq^{n-2},\ cq^{n-3},\ bcq^{2n-3} \\ b/q,\ c/q,\ c/q^2,\ c/q^3,\ bcq^{n-3} \end{bmatrix}$ $\times \left\{ 1 + cq^{n-2} \begin{bmatrix} q^n,\ b/q,\ -cq^{n-2} \\ cq^{2n-1},\ bcq^{2n-3} \end{bmatrix} \right\}$	$B: \lambda = 2$ $\mu = 2$ $\nu = 4$
0	-2	0	$q^{-2n} \begin{bmatrix} cq^{2n+2},\ bcq^{2n+1} \\ cq^{n+2},\ bcq^{n+1} \end{bmatrix}$ $\times \left\{ 1 + q^{n+1} \begin{bmatrix} q^n,\ b,\ -cq^{n+1} \\ cq^{2n+2},\ bcq^{2n+1} \end{bmatrix} \right\}$	$A: \lambda = 0$ $\mu = 0$ $\nu = 0$
		-1	$q^{-2n} \begin{bmatrix} cq^{2n+2},\ bcq^{2n+1} \\ cq^{n+2},\ bcq^{n+1} \end{bmatrix}$ $\times \left\{ 1 + cq^{n+1} \begin{bmatrix} q^n,\ b,\ -cq^{n+1} \\ cq^{2n+2},\ bcq^{2n+1} \end{bmatrix} \right\}$	

α	γ	ε	$\chi_0[\alpha,\gamma;\varepsilon]$	note
3	4	4	$q^{-2n}\begin{bmatrix} bq^{n-1},\, bq^{n-2},\, cq^{3n-1},\, cq^{n-2},\, cq^{n-3} \\ b/q,\, b/q^2,\, c/q,\, c/q^2,\, c/q^3 \end{bmatrix}$ $\times\left\{1 + q^{n+1}\begin{bmatrix} q^n,\, bcq^{n-4} \\ bq^{n-1},\, cq^{3n-1} \end{bmatrix}\right\}$	$A:\lambda=1$ $\mu=3$ $\nu=4$ $B:\lambda=2$ $\mu=3$ $\nu=4$
-1	-2	-1	$q^{-2n}\begin{bmatrix} cq^{3n+2} \\ cq^{n+2} \end{bmatrix} \times\left\{1 + q^{n+1}\begin{bmatrix} q^n,\, bcq^{n+1} \\ bq^{n+1},\, cq^{3n+2} \end{bmatrix}\right\}$	$A:\lambda=-1$ $\mu=-1$ $\nu=-1$
2	5	4	$q^{-3n}\begin{bmatrix} bq^{n-1},cq^{2n-1},cq^{2n-2},cq^{n-3},cq^{n-4},bcq^{2n-4} \\ b/q,\, c/q,\, c/q^2,\, c/q^3,\, c/q^4,\, bcq^{n-4} \end{bmatrix}$ $\times\left\{1 + q^{n+1}\begin{bmatrix} q^n,\, b/q,\, -cq^{n-3},\, c^2q^{2n-5} \\ cq^{2n-1},\, cq^{2n-2},\, bcq^{2n-4} \end{bmatrix}\right\}$	$A:\lambda=1$ $\mu=2$ $\nu=5$
0	-3	-1	$q^{-3n}\begin{bmatrix} cq^{2n+2},\, cq^{2n+3},\, bcq^{2n+1} \\ cq^{n+2},\, cq^{n+3},\, bcq^{n+1} \end{bmatrix}$ $\times\left\{1 + q^{n+1}\begin{bmatrix} q^n,\, b,\, -cq^{n+1},\, c^2q^{2n+3} \\ cq^{2n+2},\, cq^{2n+3},\, bcq^{2n+1} \end{bmatrix}\right\}$	$A:\lambda=0$ $\mu=0$ $\nu=0$
4	4	4	$q^{2-2n}\begin{bmatrix} bq^{n-2},\, bq^{n-3},\, cq^{n-2},\, cq^{n-3} \\ b/q,\, b/q^2,\, c/q,\, c/q^2 \end{bmatrix}$ $\times\left\{1 + q^{-2n}\begin{bmatrix} q^{2n+1},\, -q^{n+1},\, bcq^{n-4} \\ b/q^3,\, c/q^3 \end{bmatrix}\right\}$	$A:\lambda=1$ $\mu=4$ $\nu=4$
		5	$q^{2+2n}\begin{bmatrix} bq^{n-2},\, bq^{n-3},\, cq^{n-2},\, cq^{n-3} \\ b/q,\, b/q^2,\, c/q,\, c/q^2 \end{bmatrix}$ $\times\left\{1 + q^{-2-3n}\begin{bmatrix} q^{2n+1},\, -q^{n+1},\, bcq^{n-4} \\ b/q^3,\, c/q^3 \end{bmatrix}\right\}$	
-2	-2	-1	$q^{2-n}\begin{bmatrix} b,\, c \\ bq^{n+2},\, cq^{n+2} \end{bmatrix}$ $\times\left\{1 + q^{-2-n}\begin{bmatrix} q^{2n+1},\, -q^{n+1},\, bcq^{n+2} \\ b,\, c \end{bmatrix}\right\}$	$A:\lambda=-2$ $\mu=-2$ $\nu=-2$
		-2	$q^{2+n}\begin{bmatrix} b,\, c \\ bq^{n+2},\, cq^{n+2} \end{bmatrix}$ $\times\left\{1 + q^{-2-3n}\begin{bmatrix} q^{2n+1},\, -q^{n+1},\, bcq^{n+2} \\ b,\, c \end{bmatrix}\right\}$	

α	γ	ε	$\chi_0[\alpha,\gamma;\varepsilon]$	note
3	5	4	$q^{5-3n}/c \begin{bmatrix} bq^{n-2},\, cq^{n-3},\, cq^{n-4},\, cq^{2n-3},\, c^2q^{2n-5} \\ b/q,\ c/q,\ c/q^2,\ c/q^3,\ c/q^4 \end{bmatrix}$ $\times \left\{ 1 - \begin{bmatrix} c/q^3,\, c/q^4,\, bcq^{2n-4} \\ cq^{2n-3},\, bcq^{n-4},\, c^2q^{2n-5} \end{bmatrix} \right.$ $\left. + cq^{-n-5} \begin{bmatrix} q^{2n+1},\, cq^{2n-1},\, bcq^{2n-4} \\ b/q^2,\, cq^{2n-3},\, c^2q^{2n-5} \end{bmatrix} \right\}$	$A:\ \lambda=1$ $\mu=3$ $\nu=5$
		5	$q^{1-n} \begin{bmatrix} bq^{n-2},\, cq^{n-3},\, cq^{n-4},\, cq^{2n-3},\, c^2q^{2n-5} \\ b/q,\ c/q,\ c/q^2,\ c/q^3,\ c/q^4 \end{bmatrix}$ $\times \left\{ 1 - cq^{3n-1} \begin{bmatrix} c/q^3,\, c/q^4,\, bcq^{2n-4} \\ cq^{2n-3},\, bcq^{n-4},\, c^2q^{2n-5} \end{bmatrix} \right.$ $\left. + q^{-n-1} \begin{bmatrix} q^{2n+1},\, cq^{2n-1},\, bcq^{2n-4} \\ b/q^2,\, cq^{2n-3},\, c^2q^{2n-5} \end{bmatrix} \right\}$	
-1	-3	-1	$q^{1-2n}/c \begin{bmatrix} b,\, cq^{2n+1},\, c^2q^{2n+3} \\ bq^{n+1},\, cq^{n+2},\, cq^{n+3} \end{bmatrix}$ $\times \left\{ 1 - \begin{bmatrix} c,\, cq,\, bcq^{2n+2} \\ cq^{2n+1},\, bcq^{n+2},\, c^2q^{2n+3} \end{bmatrix} \right.$ $\left. + cq^{-n-1} \begin{bmatrix} q^{2n+1},\, cq^{2n+3},\, bcq^{2n+2} \\ b,\, cq^{2n+1},\, c^2q^{2n+3} \end{bmatrix} \right\}$	$A:\ \lambda=-1$ $\mu=-1$ $\nu=-1$
		-2	$q^{1-2n} \begin{bmatrix} b,\, cq^{2n+1},\, c^2q^{2n+3} \\ bq^{n+1},\, cq^{n+2},\, cq^{n+3} \end{bmatrix}$ $\times \left\{ 1 - cq^{3n+3} \begin{bmatrix} c,\, cq,\, bcq^{2n+2} \\ cq^{2n+1},\, bcq^{n+2},\, c^2q^{2n+3} \end{bmatrix} \right.$ $\left. + q^{-n-1} \begin{bmatrix} q^{2n+1},\, cq^{2n+3},\, bcq^{2n+2} \\ b,\, cq^{2n+1},\, c^2q^{2n+3} \end{bmatrix} \right\}$	
4	5	5	$q^{2-2n} \begin{bmatrix} bq^{n-2},\, bq^{n-3},\, cq^{n-3},\, cq^{n-4},\, cq^{4n-1} \\ b/q,\ b/q^2,\ c/q,\ c/q^2,\ c/q^3 \end{bmatrix}$ $\times \left\{ 1 + q^{-2-n} \begin{bmatrix} q^{2n+1},\, cq^{2n-1},\, -q^{n+1},\, bcq^{n-5} \\ b/q^3,\, c/q^4,\, cq^{4n-1} \end{bmatrix} \right\}$	$A:\ \lambda=1$ $\mu=4$ $\nu=5$
-2	-3	-2	$q^{2-2n} \begin{bmatrix} b,\, c,\, cq^{4n+3} \\ bq^{n+2},\, cq^{n+2},\, cq^{n+3} \end{bmatrix}$ $\times \left\{ 1 + q^{-2-n} \begin{bmatrix} q^{2n+1},\, cq^{2n+3},\, -q^{n+1},\, bcq^{n+2} \\ b,\, c,\, cq^{4n+3} \end{bmatrix} \right\}$	$A:\ \lambda=-2$ $\mu=-2$ $\nu=-2$

1.3 Basic almost poised formulas when $\delta = 1$

α	γ	ε	$\chi_1[\alpha, \gamma; \varepsilon]$	note
0	0	-1	$\left[q^{-n-1}, -q^{-2n-1}\right] \times \left\{1 + q^{n+1}\begin{bmatrix} q^n, b, c \\ bcq^{n+1}, -q^{2n+1} \end{bmatrix}\right\}$	$A: \lambda = 0$ $\mu = 0$ $\nu = 0$
		0	$1 - q^{-n-1}$	
		1	0	
		2	$q^n \times (1 - q^{n+1})$	
		3	$q^n\left[q^{n+1}, -q^{2n+1}\right] \times \left\{1 + q^{n+1}\begin{bmatrix} q^n, b, c \\ bcq^{n+1}, -q^{2n+1} \end{bmatrix}\right\}$	(cf. [5, 16])
1	-1	0	$q^{-2n}/b\begin{bmatrix} q^{n+1}, qc/b \\ cq^{n+1}, q/b \end{bmatrix} \times \left\{1 - c\begin{bmatrix} q/b, bq^n, bcq^{2n} \\ qc/b, bcq^n \end{bmatrix}\right\}$	$A: \lambda = 0$ $\mu = 1$ $\nu = 0$
		1	$\begin{bmatrix} q^{n+1}, qc/b \\ cq^{n+1}, q/b \end{bmatrix}$	
		2	$bq^{2n}\begin{bmatrix} q^{n+1}, qc/b \\ cq^{n+1}, q/b \end{bmatrix} \times \left\{1 + q^{-2n}/b\begin{bmatrix} q/b, bq^n, bcq^{2n} \\ qc/b, bcq^n \end{bmatrix}\right\}$	$D: \lambda = 0$ $\mu = 0$ $\nu = 0$
2	-2	1	$q^{-n}\begin{bmatrix} q^{n+1}, q^2c/b, bq^{n-1}, -cq^n \\ q/b, cq^{n+2}, bcq^{n-1} \end{bmatrix}$ $\times \left\{1 - q/b\begin{bmatrix} cq, -bq^{n-1} \\ q^2/b, -cq^n \end{bmatrix}\right\}$	$A: \lambda = 0$ $\mu = 2$ $\nu = 0$ $D: \lambda = 0$ $\mu = 1$ $\nu = 0$

α	γ	ε	$\chi_1[\alpha,\gamma;\varepsilon]$	note
0	1	0	$q^{-2n}/c \begin{bmatrix} q^{-n-1} \\ c/q \end{bmatrix} \times \left\{ 1 - \begin{bmatrix} c/q,\, cq^n,\, bcq^{2n} \\ bcq^n \end{bmatrix} \right\}$	$A: \lambda=0$
		1	$\begin{bmatrix} q^{-n-1} \\ c/q \end{bmatrix}$	$\mu=0$
		2	$cq^{2n} \begin{bmatrix} q^{-n-1} \\ c/q \end{bmatrix}$	$\nu=1$
		3	$c^2 q^{4n} \begin{bmatrix} q^{-n-1} \\ c/q \end{bmatrix} \times \left\{ 1 - q^{1-2n}/c^2 \begin{bmatrix} c/q,\, cq^n,\, bcq^{2n} \\ bcq^n \end{bmatrix} \right\}$	
0	-1	-1	$c^2 \begin{bmatrix} q^{n+1} \\ cq^{n+1} \end{bmatrix} \times \left\{ 1 - q^{-1-2n}/c^2 \begin{bmatrix} c,\, cq^{n+1},\, bcq^{2n+1} \\ bcq^{n+1} \end{bmatrix} \right\}$	$A: \lambda=0$
		0	$c \begin{bmatrix} q^{n+1} \\ cq^{n+1} \end{bmatrix}$	$\mu=0$
		1	$\begin{bmatrix} q^{n+1} \\ cq^{n+1} \end{bmatrix}$	$\nu=0$
		2	$c^{-1} \begin{bmatrix} q^{n+1} \\ cq^{n+1} \end{bmatrix} \times \left\{ 1 - \begin{bmatrix} c,\, cq^{n+1},\, bcq^{2n+1} \\ bcq^{n+1} \end{bmatrix} \right\}$	
1	1	1	$\begin{bmatrix} q^{-n-1} \\ c/q \end{bmatrix} \times \left\{ 1 + q^{-n-1} \begin{bmatrix} cq^n \\ b/q \end{bmatrix} \right\}$	$A: \lambda=0$
		2	$q^n \begin{bmatrix} q^{-n-1},\, bcq^{n-1} \\ b/q,\, c/q \end{bmatrix}$	$\mu=1$
		3	$cq^{3n} \begin{bmatrix} q^{-n-1} \\ c/q \end{bmatrix} \times \left\{ 1 + b/c \begin{bmatrix} cq^n \\ b/q \end{bmatrix} \right\}$	$\nu=1$
-1	-1	-1	$\begin{bmatrix} q^{n+1},\, bcq^{n+1} \\ bq^{n+1},\, cq^{n+1} \end{bmatrix} \times \left\{ 1 - \begin{bmatrix} b,\, c \\ bcq^{n+1} \end{bmatrix} \right\}$	$A: \lambda=-1$
		0	$\begin{bmatrix} q^{n+1},\, bcq^{n+1} \\ bq^{n+1},\, cq^{n+1} \end{bmatrix}$	$\mu=-1$
		1	$\begin{bmatrix} q^{n+1},\, bcq^{n+1} \\ bq^{n+1},\, cq^{n+1} \end{bmatrix} \times \left\{ 1 + q^{n+1} \begin{bmatrix} b,\, c \\ bcq^{n+1} \end{bmatrix} \right\}$	$\nu=-1$

α	γ	ε	$\chi_1[\alpha,\gamma;\varepsilon]$	note
0	2	1	$q^{-n}\begin{bmatrix} q^{-2n-2},\ cq^{n-1} \\ c/q,\ c/q^2 \end{bmatrix} \times \left\{ 1 - bq^{n+1}\begin{bmatrix} q^n,\ c/q^2 \\ bcq^{n-1},\ -q^{n+1} \end{bmatrix} \right\}$	$A:\ \lambda=0$ $\mu=0$ $\nu=2$
		2	$\begin{bmatrix} q^{-n-1},\ c^2q^{2n-2} \\ c/q,\ c/q^2 \end{bmatrix}$	
		3	$cq^{3n}\begin{bmatrix} q^{-2n-2},\ cq^{n-1} \\ c/q,\ c/q^2 \end{bmatrix} \times \left\{ 1 + q\begin{bmatrix} q^n,\ c/q^2 \\ bcq^{n-1},\ -q^{n+1} \end{bmatrix} \right\}$	
0	-2	-1	$cq^{-n}\begin{bmatrix} q^{2n+2} \\ cq^{n+2} \end{bmatrix} \times \left\{ 1 + q\begin{bmatrix} q^n,\ c \\ bcq^{n+1},\ -q^{n+1} \end{bmatrix} \right\}$	$A:\ \lambda=0$ $\mu=0$ $\nu=0$
		0	$q^{-n}\begin{bmatrix} q^{n+1},\ -cq^{n+1} \\ cq^{n+2} \end{bmatrix}$	
		1	$q^{-n}\begin{bmatrix} q^{2n+2} \\ cq^{n+2} \end{bmatrix} \times \left\{ 1 - bq^{n+1}\begin{bmatrix} q^n,\ c \\ bcq^{n+1},\ -q^{n+1} \end{bmatrix} \right\}$	
1	2	2	$\begin{bmatrix} q^{-n-1},\ cq^{n-1} \\ c/q,\ c/q^2 \end{bmatrix} \times \left\{ 1 + q^{-n-1}\begin{bmatrix} bcq^{2n-1} \\ b/q \end{bmatrix} \right\}$	$A:\ \lambda=0$ $\mu=1$ $\nu=2$
		3	$cq^{2n-1}\begin{bmatrix} q^{-n-1},\ cq^{n-1} \\ c/q,\ c/q^2 \end{bmatrix} \times \left\{ 1 + q^{1-n}/c\begin{bmatrix} bcq^{2n-1} \\ b/q \end{bmatrix} \right\}$	
-1	-2	0	$q^{-n}\begin{bmatrix} q^{n+1} \\ cq^{n+2} \end{bmatrix} \times \left\{ 1 + q^{n+1}\begin{bmatrix} bcq^{n+1} \\ bq^{n+1} \end{bmatrix} \right\}$	$A:\ \lambda=-1$ $\mu=-1$ $\nu=-1$
		-1	$cq\begin{bmatrix} q^{n+1} \\ cq^{n+2} \end{bmatrix} \times \left\{ 1 + q^{-n-1}/c\begin{bmatrix} bcq^{n+1} \\ bq^{n+1} \end{bmatrix} \right\}$	

α	γ	ε	$\chi_1[\alpha, \gamma; \varepsilon]$	note
0	3	2	$q^{2-2n}/c \begin{bmatrix} q^{-n-1}, c^2q^{2n-3}, c^2q^{2n-4} \\ c/q, c/q^2, c/q^3 \end{bmatrix}$ $\times \left\{ 1 - \begin{bmatrix} c/q^3, bcq^{2n-2} \\ bcq^{n-2}, -cq^{n-2}, c^2q^{2n-3} \end{bmatrix} \right\}$	$A: \lambda = 0$ $\mu = 0$ $\nu = 3$
		3	$\begin{bmatrix} q^{-n-1}, c^2q^{2n-3}, c^2q^{2n-4} \\ c/q, c/q^2, c/q^3 \end{bmatrix} \times$ $\left\{ 1 - c^2q^{2n-2} \begin{bmatrix} c/q^3, bcq^{2n-2} \\ bcq^{n-2}, -cq^{n-2}, c^2q^{2n-3} \end{bmatrix} \right\}$	
0	-3	0	$q^{-2n}/c \begin{bmatrix} q^{n+1}, -cq^{n+1}, c^2q^{2n+3} \\ cq^{n+2}, cq^{n+3} \end{bmatrix}$ $\times \left\{ 1 - \begin{bmatrix} c, bcq^{2n+1} \\ bcq^{n+1}, -cq^{n+1}, c^2q^{2n+3} \end{bmatrix} \right\}$	$A: \lambda = 0$ $\mu = 0$ $\nu = 0$
		-1	$q^{-2n} \begin{bmatrix} q^{n+1}, -cq^{n+1}, c^2q^{2n+3} \\ cq^{n+2}, cq^{n+3} \end{bmatrix}$ $\times \left\{ 1 - c^2q^{2n+4} \begin{bmatrix} c, bcq^{2n+1} \\ bcq^{n+1}, -cq^{n+1}, c^2q^{2n+3} \end{bmatrix} \right\}$	
1	-3	0	$b \begin{bmatrix} q^{n+1}, c, q^2c/b, q^3c/b \\ q/b, cq^{n+2}, cq^{n+3}, bcq^n \end{bmatrix}$ $\times \left\{ 1 + q^{-2n}/b \begin{bmatrix} q/b, bq^n, -cq^{n+1}, c^2q^{2n+3} \\ c, q^2c/b, q^3c/b \end{bmatrix} \right\}$	$A: \lambda = 0$ $\mu = 1$ $\nu = 0$ $D: \lambda = 0$ $\mu = 0$ $\nu = 0$
-1	3	2	$\begin{bmatrix} q^{n+1}, qb/c, q^2b/c, cq^{n-2} \\ q/c, q^2/c, bq^{n+1}, bcq^{n-2} \end{bmatrix} \times$ $\left\{ 1 + q^{5-2n}c^{-3} \begin{bmatrix} b, bq^{n+1}, -cq^{n-2}, c^2q^{2n-3} \\ qb/c, q^2b/c, q^3/c \end{bmatrix} \right\}$	$A: \lambda = 0$ $\mu = 0$ $\nu = 3$

α	γ	ε	$\chi_1[\alpha,\gamma;\varepsilon]$	note
2	2	2	$q^{-n}\begin{bmatrix} q^{-n-1},\ cq^{n-1},\ b^2q^{2n-2} \\ b/q,\ c/q,\ b/q^2 \end{bmatrix}$ $\times\left\{1-q^{-n}/c\begin{bmatrix} bq^n,\ -q^{n+1} \\ q^2/c,\ -bq^{n-1} \end{bmatrix}\right\}$	$A:\lambda=0$ $\mu=2$ $\nu=2$
		3	$q^n\begin{bmatrix} q^{-2n-2},\ bq^{n-1},\ cq^{n-1},\ bcq^{n-2} \\ b/q,\ b/q^2,\ c/q,\ c/q^2 \end{bmatrix}$	
		4	$q^{2n}\begin{bmatrix} q^{-n-1},\ cq^{n-1},\ b^2q^{2n-2} \\ b/q,\ c/q,\ b/q^2 \end{bmatrix}$ $\times\left\{1-\begin{bmatrix} bq^n,\ -q^{n+1} \\ q^2/c,\ -bq^{n-1} \end{bmatrix}\right\}$	
-2	-2	0	$q^{-n}\begin{bmatrix} q^{2n+2},\ bcq^{n+2} \\ bq^{n+2},\ cq^{n+2} \end{bmatrix}$ $\times\left\{1+q^{n+2}\begin{bmatrix} b,\ c \\ bcq^{n+2},\ -q^{n+1} \end{bmatrix}\right\}$	$A:\lambda=-2$ $\mu=-2$ $\nu=-2$
		-1	$q^{-n}\begin{bmatrix} q^{2n+2},\ bcq^{n+2} \\ bq^{n+2},\ cq^{n+2} \end{bmatrix}$	
		-2	$q^{-n}\begin{bmatrix} q^{2n+2},\ bcq^{n+2} \\ bq^{n+2},\ cq^{n+2} \end{bmatrix}$ $\times\left\{1-q^{n+1}\begin{bmatrix} b,\ c \\ bcq^{n+2},\ -q^{n+1} \end{bmatrix}\right\}$	
1	3	3	$\begin{bmatrix} q^{-n-1},\ cq^{n-1},\ cq^{n-2},\ bcq^{n-3} \\ b/q,\ c/q,\ c/q^2,\ c/q^3 \end{bmatrix}\times$ $\left\{1+q^{-n-1}\begin{bmatrix} bq^n,\ c^2q^{3n-2} \\ cq^{n-1},\ bcq^{n-3} \end{bmatrix}\right\}$	$A:\lambda=0$ $\mu=1$ $\nu=3$
-1	-3	-1	$q^{1-n}\begin{bmatrix} q^{n+1},\ bcq^{n+1} \\ bq^{n+1},\ cq^{n+3} \end{bmatrix}$ $\times\left\{1+q^{-n-1}\begin{bmatrix} bq^{n+1},\ c^2q^{3n+4} \\ cq^{n+2},\ bcq^{n+1} \end{bmatrix}\right\}$	$A:\lambda=-1$ $\mu=-1$ $\nu=-1$

α	γ	ε	$\chi_1[\alpha,\gamma;\varepsilon]$	note
2	3	3	$q^{-n}\begin{bmatrix} q^{-n-1},\,bq^{n-1},\,cq^{n-2} \\ b/q,\,c/q,\,c/q^2 \end{bmatrix}$ $\times\left\{1+q^{-n-2}\begin{bmatrix} cq^{2n},\,-q^{n+1},\,bcq^{n-3} \\ b/q^2,\,c/q^3 \end{bmatrix}\right\}$	$A:\lambda=0$ $\mu=2$ $\nu=3$
		4	$cq^{3n}\begin{bmatrix} q^{-n-1},\,bq^{n-1},\,cq^{n-2} \\ b/q,\,c/q,\,c/q^2 \end{bmatrix}$ $\times\left\{1+q^{-3n-1}/c\begin{bmatrix} cq^{2n},\,-q^{n+1},\,bcq^{n-3} \\ b/q^2,\,c/q^3 \end{bmatrix}\right\}$	
-2	-3	-1	$q^{2-n}\begin{bmatrix} q^{n+1},\,b,\,c \\ bq^{n+2},\,cq^{n+2},\,cq^{n+3} \end{bmatrix}$ $\times\left\{1+q^{-n-2}\begin{bmatrix} cq^{2n+3},\,-q^{n+1},\,bcq^{n+2} \\ b,\,c \end{bmatrix}\right\}$	$A:\lambda=-2$ $\mu=-2$ $\nu=-2$
		-2	$cq^{4+n}\begin{bmatrix} q^{n+1},\,b,\,c \\ bq^{n+2},\,cq^{n+2},\,cq^{n+3} \end{bmatrix}$ $\times\left\{1+q^{-3n-4}/c\begin{bmatrix} cq^{2n+3},\,-q^{n+1},\,bcq^{n+2} \\ b,\,c \end{bmatrix}\right\}$	
3	3	4	$q^{n}\begin{bmatrix} q^{-3n-3},\,bq^{n-1},\,bq^{n-2},\,cq^{n-1},\,cq^{n-2},\,bcq^{n-3} \\ b/q,\,b/q^2,\,b/q^3,\,c/q,\,c/q^2,\,c/q^3 \end{bmatrix}$ $\times\left\{1+q^{n+2}\begin{bmatrix} q^n,\,q^{n+1},\,bcq^{n-4} \\ bq^{n-1},\,cq^{n-1},\,q^{3n+3} \end{bmatrix}\right\}$	$A:\lambda=0$ $\mu=3$ $\nu=3$
-3	-3	-2	$q^{-2n}\begin{bmatrix} q^{3n+3},\,bcq^{n+3} \\ bq^{n+3},\,cq^{n+3} \end{bmatrix}$ $\times\left\{1+q^{n+2}\begin{bmatrix} q^n,\,q^{n+1},\,bcq^{n+2} \\ q^{3n+3},\,bq^{n+2},\,cq^{n+2} \end{bmatrix}\right\}$	$A:\lambda=-3$ $\mu=-3$ $\nu=-3$
2	4	4	$q^{-2-2n}\begin{bmatrix} q^{-n-1},\,bq^n,\,bq^{n-1},\,cq^{n-2},\,cq^{n-3},\,c^2q^{4n-2} \\ b/q,\,b/q^2,\,c/q,\,c/q^2,\,c/q^3,\,c/q^4 \end{bmatrix}$ $\times\left\{1+q^{n+1}\begin{bmatrix} -q,\,bcq^{n-4} \\ bq^n,\,-cq^{2n-1} \end{bmatrix}\right\}$	$A:\lambda=0$ $\mu=2$ $\nu=4$
-2	-4	-2	$q^{-3n}\begin{bmatrix} q^{n+1},\,c^2q^{4n+6} \\ cq^{n+3},\,cq^{n+4} \end{bmatrix}$ $\times\left\{1+q^{n+1}\begin{bmatrix} -q,\,bcq^{n+2} \\ bq^{n+2},\,-cq^{2n+3} \end{bmatrix}\right\}$	$A:\lambda=-2$ $\mu=-2$ $\nu=-2$

α	γ	ε	$\chi_1[\alpha,\gamma;\varepsilon]$	note
3	4	4	$q^{-3-3n}\left[\begin{matrix}q^{-n-1},\,bq^n,\,bq^{n-1},\,bq^{n-2},\,cq^{n-1},\,cq^{n-2},\,cq^{n-3}\\[2pt] b/q,\,b/q^2,\,b/q^3,\,c/q,\,c/q^2,\,c/q^3,\,c/q^4\end{matrix}\right]$ $\times\left\{1+q^{n+3}\left[\begin{matrix}cq^{3n-1},\ bcq^{n-4}\\ bq^n,\ cq^{n-1}\end{matrix}\right]\right.$ $+q^{n+1}\left[\begin{matrix}-q,\,cq^{2n-1},\,bcq^{n-4}\\ bq^n,\,cq^{n-1}\end{matrix}\right]$ $\left.+q^{2n+4}\left[\begin{matrix}q^n,\,c/q^4,\,bcq^{n-4}\\ bq^n,\,bq^{n-1},\,cq^{n-1}\end{matrix}\right]\right\}$	$A:\ \lambda=0$ $\mu=3$ $\nu=4$
		5	$cq^{-3+2n}\left[\begin{matrix}q^{-n-1},\,bq^n,\,bq^{n-1},\,bq^{n-2},\,cq^{n-1},\,cq^{n-2},\,cq^{n-3}\\[2pt] b/q,\,b/q^2,\,b/q^3,\,c/q,\,c/q^2,\,c/q^3,\,c/q^4\end{matrix}\right]$ $\times\left\{1+q^{1-3n}/c\left[\begin{matrix}cq^{3n-1},\ bcq^{n-4}\\ bq^n,\ cq^{n-1}\end{matrix}\right]\right.$ $+q^{2-2n}/c\left[\begin{matrix}-q,\,cq^{2n-1},\,bcq^{n-4}\\ bq^n,\,cq^{n-1}\end{matrix}\right]$ $\left.+q^{2-n}b/c\left[\begin{matrix}q^n,\,c/q^4,\,bcq^{n-4}\\ bq^n,\,bq^{n-1},\,cq^{n-1}\end{matrix}\right]\right\}$	
-3	-4	-2	$q^{-3n}\left[\begin{matrix}q^{n+1}\\ cq^{n+4}\end{matrix}\right]$ $\times\left\{1+q^{n+3}\left[\begin{matrix}cq^{3n+3},\ bcq^{n+3}\\ bq^{n+3},\ cq^{n+3}\end{matrix}\right]\right.$ $+q^{n+1}\left[\begin{matrix}-q,\,cq^{2n+3},\,bcq^{n+3}\\ bq^{n+3},\,cq^{n+3}\end{matrix}\right]$ $\left.+q^{2n+4}\left[\begin{matrix}q^n,\,c,\,bcq^{n+3}\\ bq^{n+2},\,bq^{n+3},\,cq^{n+3}\end{matrix}\right]\right\}$	$A:\ \lambda=-3$ $\mu=-3$ $\nu=-3$
		-3	$cq^3\left[\begin{matrix}q^{n+1}\\ cq^{n+4}\end{matrix}\right]$ $\times\left\{1+q^{-3-3n}/c\left[\begin{matrix}cq^{3n+3},\ bcq^{n+3}\\ bq^{n+3},\ cq^{n+3}\end{matrix}\right]\right.$ $+q^{-2-2n}/c\left[\begin{matrix}-q,\,cq^{2n+3},\,bcq^{n+3}\\ bq^{n+3},\,cq^{n+3}\end{matrix}\right]$ $\left.+q^{1-n}b/c\left[\begin{matrix}q^n,\,c,\,bcq^{n+3}\\ bq^{n+2},\,bq^{n+3},\,cq^{n+3}\end{matrix}\right]\right\}$	

α	γ	ε	$\chi_1[\alpha, \gamma; \varepsilon]$	note
4	4	5	$q^{-4-3n}\begin{bmatrix} q^{4n+4}, bq^{n-1}, bq^{n-2}, bq^{n-3} \\ b/q, b/q^2, b/q^3, b/q^4 \end{bmatrix}$ $\times \begin{bmatrix} cq^{n-1}, cq^{n-2}, cq^{n-3}, bcq^{n-4} \\ c/q, c/q^2, c/q^3, c/q^4 \end{bmatrix}$ $\times \left\{ 1 + q^{n+2}\begin{bmatrix} q^n, -q, bcq^{n-5} \\ bq^{n-1}, cq^{n-1}, -q^{2n+2} \end{bmatrix} \right\}$	$A: \lambda = 0$ $\mu = 4$ $\nu = 4$
-4	-4	-3	$q^{-3n}\begin{bmatrix} q^{4n+4}, bcq^{n+4} \\ bq^{n+4}, cq^{n+4} \end{bmatrix}$ $\times \left\{ 1 + q^{n+2}\begin{bmatrix} q^n, -q, bcq^{n+3} \\ bq^{n+3}, cq^{n+3}, -q^{2n+2} \end{bmatrix} \right\}$	$A: \lambda = -4$ $\mu = -4$ $\nu = -4$
1	-2	0	$q^{-n}\begin{bmatrix} q^{n+1}, qc/b, -cq^{n+1} \\ q/b, cq^{n+2} \end{bmatrix}$ $\times \left\{ 1 - q^2 c/b \begin{bmatrix} c, bcq^{2n} \\ qc/b, -cq^{n+1}, bcq^n \end{bmatrix} \right\}$	$A: \lambda = 0$ $\mu = 1$ $\nu = 0$
		1	$q^{-n}/c \begin{bmatrix} q^{n+1}, qc/b, -cq^{n+1} \\ q/b, cq^{n+2} \end{bmatrix}$ $\times \left\{ 1 - \begin{bmatrix} c, bcq^{2n} \\ qc/b, -cq^{n+1}, bcq^n \end{bmatrix} \right\}$	$D: \lambda = 0$ $\mu = 0$ $\nu = 0$
-1	2	1	$q^{-n}\begin{bmatrix} q^{-n-1}, q^2 b/c, c^2 q^{2n-2} \\ c/q, c/q^2, bq^{n+1} \end{bmatrix}$ $\times \left\{ 1 - b \begin{bmatrix} q^2/c, bcq^{2n-1} \\ q^2 b/c, -cq^{n-1}, bcq^{n-1} \end{bmatrix} \right\}$	$A: \lambda = 0$ $\mu = 0$ $\nu = 2$
		2	$cq^{n-1}\begin{bmatrix} q^{-n-1}, q^2 b/c, c^2 q^{2n-2} \\ c/q, c/q^2, bq^{n+1} \end{bmatrix}$ $\times \left\{ 1 - c/q \begin{bmatrix} q^2/c, bcq^{2n-1} \\ q^2 b/c, -cq^{n-1}, bcq^{n-1} \end{bmatrix} \right\}$	

1.4 Additional formulae for $_4\phi_3[\cdots]$ and $_5\phi_4[\cdots]$

Recall the linear combination

$$
_4\phi_3\left[\begin{array}{cccc} q^{-\delta-2n}, & qw, & b, & c \\ & w, & q^{\alpha-2n}/b, & q^{\gamma-2n}/c \end{array}; q^{\varepsilon-n}/bc\right]
$$

$$
= \frac{1}{1-w}\,\omega_\delta\left[\begin{array}{cc} \alpha, & \gamma; & \varepsilon \\ b, & c; & n \end{array}\right] - \frac{w}{1-w}\,\omega_\delta\left[\begin{array}{cc} \alpha, & \gamma; & 1+\varepsilon \\ b, & c; & n \end{array}\right]
$$

$$
= \frac{1}{1-w}\left\{\chi_\delta\left[\begin{array}{c} \alpha,\gamma;\varepsilon \\ b,c;n \end{array}\right] - w\,\chi_\delta\left[\begin{array}{c} \alpha,\gamma;1+\varepsilon \\ b,c;\ n \end{array}\right]\right\}\left[\begin{array}{cc} q^{-\delta-2n}, & q^{[\alpha,\gamma]-2n}/bc \\ q^{\alpha-2n}/b, & q^{\gamma-2n}/c \end{array}; q\right]_n.
$$

Based on the tables exhibited in the last two sections, a number of evaluations may be established. Among them, the simple shifted product evaluations read as follows.

Example 1.1

$$
_4\phi_3\left[\begin{array}{cccc} q^{-2n}, & qw, & b, & c \\ & w, & q^{-2n}/b, & q^{-2n}/c \end{array}; q^{-n}/bc\right]
$$

$$
= \left[\begin{array}{cc} q^{-2n}, & q^{-2n}/bc \\ q^{-2n}/b, & q^{-2n}/c \end{array}; q\right]_n.
$$

Example 1.2

$$
_4\phi_3\left[\begin{array}{cccc} q^{-2n}, & qw, & b, & c \\ & w, & q^{1-2n}/b, & q^{1-2n}/c \end{array}; q^{1-n}/bc\right]
$$

$$
= \left[\begin{array}{c} wq^n \\ w \end{array}\right] \times \left[\begin{array}{cc} q^{-2n}, & q^{1-2n}/bc \\ q^{1-2n}/b, & q^{1-2n}/c \end{array}; q\right]_n.
$$

Example 1.3

$$
_4\phi_3\left[\begin{array}{cccc} q^{-2n}, & qw, & b, & c \\ & w, & q^{2-2n}/b, & q^{2-2n}/c \end{array}; q^{2-n}/bc\right]
$$

$$
= \left[\begin{array}{ccc} bq^{n-1}, & cq^{n-1}, & wq^{2n} \\ b/q, & c/q, & w \end{array}\right] \times \left[\begin{array}{cc} q^{-2n}, & q^{2-2n}/bc \\ q^{2-2n}/b, & q^{2-2n}/c \end{array}; q\right]_n.
$$

Example 1.4

$$
_4\phi_3\left[\begin{array}{cccc} q^{-1-2n}, & qw, & b, & c \\ & w, & q^{-2n}/b, & q^{-2n}/c \end{array}; q^{\varepsilon-n}/bc\right]
$$

$$
\overset{\varepsilon=0,1}{=\!=\!=}\ (q^{1+2n}w)^\varepsilon\left[\begin{array}{c} q^{-n-1} \\ w \end{array}\right] \times \left[\begin{array}{cc} q^{-1-2n}, & q^{-2n}/bc \\ q^{-2n}/b, & q^{-2n}/c \end{array}; q\right]_n.
$$

Example 1.5

$$
_4\phi_3\left[\begin{array}{cccc} q^{-1-2n}, & qw, & b, & c \\ & w, & q^{-2n}/b, & q^{1-2n}/c \end{array}; q^{1-n}/bc\right]
$$

$$
= \left[\begin{array}{cc} q^{-n-1}, & cwq^{2n} \\ c/q, & w \end{array}\right] \times \left[\begin{array}{cc} q^{-1-2n}, & q^{1-2n}/bc \\ q^{-2n}/b, & q^{1-2n}/c \end{array}; q\right]_n.
$$

Example 1.6

$$
{}_4\phi_3 \begin{bmatrix} q^{-1-2n}, & qw, & b, & c \\ & w, & q^{-2n}/b, & q^{-1-2n}/c \end{bmatrix} ; q^{-n}/bc \end{bmatrix}
$$

$$
= \begin{bmatrix} q^{-n-1}, & w/c \\ q^{-n-1}/c, & w \end{bmatrix} \times \begin{bmatrix} q^{-1-2n}, & q^{-2n}/bc \\ q^{-2n}/b, & q^{-1-2n}/c \end{bmatrix} ; q \end{bmatrix}_n .
$$

Similarly, from

$$
{}_5\phi_4 \begin{bmatrix} q^{-\delta-2n}, qu, qv, & b, & c \\ u, & v, q^{\alpha-2n}/b, q^{\gamma-2n}/c \end{bmatrix} ; q^{\varepsilon-n}/bc \end{bmatrix} = \frac{1}{(1-u)(1-v)}
$$

$$
\times \left\{ \omega_\delta \begin{bmatrix} \alpha, \gamma; \varepsilon \\ b, c; n \end{bmatrix} - (u+v) \, \omega_\delta \begin{bmatrix} \alpha, \gamma; 1+\varepsilon \\ b, c; & n \end{bmatrix} + uv \, \omega_\delta \begin{bmatrix} \alpha, \gamma; 2+\varepsilon \\ b, c; & n \end{bmatrix} \right\}
$$

we get another exceptional evaluation.

Example 1.7

$$
{}_5\phi_4 \begin{bmatrix} q^{-1-2n}, & qu, & qv, & b, & c \\ & u, & v, & q^{-2n}/b, & q^{-2n}/c \end{bmatrix} ; q^{-n}/bc \end{bmatrix}
$$

$$
= \begin{bmatrix} q^{-n-1}, uvq^{2n+1} \\ u, & v \end{bmatrix} \times \begin{bmatrix} q^{-1-2n}, & q^{-2n}/bc \\ q^{-2n}/b, & q^{-2n}/c \end{bmatrix} ; q \end{bmatrix}_n .
$$

1.5   Equivalent classes under reversal operations

Recall a couple of basic relations about q-shifted factorials

$$(u^{-1}; q^{-1})_k = (-1)^k u^{-k} q^{-\binom{k}{2}} (u; q)_k, \tag{1.17a}$$

$$(v; q)_{m-k} = (-1)^k v^{-k} \frac{(v; q)_m}{(q^{1-m}/v; q)_k} q^{-km+\binom{k+1}{2}}. \tag{1.17b}$$

The former enables us to perform an inversion $\mathcal{C}'$ on basic hypergeometric series $_{1+p}\phi_p \begin{bmatrix} a_0, a_1, \cdots, a_p \\ b_1, \cdots, b_p \end{bmatrix} ; x \end{bmatrix}$, with respect to base q, indeterminate x and parameters $\{a_i, b_j\}$ by setting $q \to 1/q$, $x \to 1/x$ and $a_i \to 1/a_i$, $b_j \to 1/b_j$. When the series is terminating, the latter can be used to reverse the summation order. For almost-poised series, its reversal can be stated explicitly as

$$\omega_\delta \begin{bmatrix} \alpha, \gamma; \varepsilon \\ b, c; n \end{bmatrix} = q^{(\delta+2n)\,(\varepsilon-\alpha-\gamma+\frac{1-\delta}{2})} \begin{bmatrix} b, & c \\ bq^{1-\alpha-\delta}, cq^{1-\gamma-\delta} & ; q \end{bmatrix}_{\delta+2n} \tag{1.18}$$
$$\times (-1)^\delta \omega_\delta \begin{bmatrix} 2-\alpha-2\delta, 2-\gamma-2\delta; 3-\varepsilon-\delta \\ bq^{1-\alpha-\delta}, \quad cq^{1-\gamma-\delta}; \quad n \end{bmatrix}$$

which will be used to determine another series-transform $\mathcal{R}'$.

After some trivial modification (replacing parameters and removing extra factors), we may define involutions on basic almost poised series by

$$\mathcal{C}' : \omega_\delta \begin{bmatrix} \alpha, \gamma; \varepsilon \\ b, c; n \end{bmatrix} \longmapsto \omega_\delta \begin{bmatrix} \alpha, & \gamma; & 1-\varepsilon+\delta+\alpha+\gamma \\ b, & c; & n \end{bmatrix}, \tag{1.19a}$$

$$\mathcal{R}' : \omega_\delta \begin{bmatrix} \alpha, \gamma; \varepsilon \\ b, c; n \end{bmatrix} \longmapsto \omega_\delta \begin{bmatrix} 2-\alpha-2\delta, 2-\gamma-2\delta; 3-\varepsilon-\delta \\ b, & c; & n \end{bmatrix}. \tag{1.19b}$$

They generate a Coxeter group with 4-elements $\mathbf{G} = \{\mathcal{I}', \mathcal{C}', \mathcal{R}', \mathcal{C}'\mathcal{R}' = \mathcal{R}'\mathcal{C}'\}$, where $\mathcal{I}'$ is an identity and

$$\mathcal{C}'\mathcal{R}' : \omega_\delta \begin{bmatrix} \alpha, \gamma; \varepsilon \\ b, c; n \end{bmatrix} \longmapsto \omega_\delta \begin{bmatrix} 2-\alpha-2\delta, 2-\gamma-2\delta; 2+\varepsilon-2\delta-\alpha-\gamma \\ b, & c; & n \end{bmatrix}. \tag{1.19c}$$

Under the action of this group $\mathbf{G}$, we can classify, according to its orbits, the almost poised formulas displayed in Section 1.2 and Section 1.3. From the subsequent tables, we conclude that we have essentially 43 different almost poised formulas.

	Equivalent classes for $\omega_0 \begin{bmatrix} \alpha, \gamma; & \varepsilon \\ b, c; & n \end{bmatrix}$

Nr	**G**-orbits: $\{(\alpha, \gamma; \varepsilon)\}$			
1	(0,2;1),	(0,2;2)		
2	(0,3;2),	(2,-1;1)		
3	(1,1;0),	(1,1;3)		
4	(1,1;1),	(1,1;2)		
5	(1,2;1),	(1,2;3),	(0,1;0),	(0,1;2)
6	(1,2;2),	(0,1;1)		
7	(1,3;2),	(1,3;3),	(1,-1;0),	(1,-1;1)
8	(1,4;3),	(1,-2;0)		
9	(2,2;1),	(2,2;4),	(0,0;-1),	(0,0;2)
10	(2,2;2),	(2,2;3),	(0,0;0),	(0,0;1)
11	(2,3;2),	(2,3;4),	(0,-1;1),	(0,-1;-1)
12	(2,3;3),	(0,-1;0)		
13	(2,4;3),	(2,4;4),	(0,-2;0),	(0,-2;-1)
14	(2,5;4),	(0,-3;-1)		
15	(3,3;3),	(3,3;4),	(-1,-1;0),	(-1,-1;-1)
16	(3,4;4),	(-1,-2;-1)		
17	(3,5;4),	(3,5;5),	(-1,-3;-1),	(-1,-3;-2)
18	(4,4;4),	(4,4;5),	(-2,-2;-1),	(-2,-2;-2)
19	(4,5;5),	(-2,-3;-2)		

Equivalent classes for $\omega_1 \begin{bmatrix} \alpha, \gamma; & \varepsilon \\ b, c; & n \end{bmatrix}$			
Nr	**G**-orbits: $\{(\alpha, \gamma; \varepsilon)\}$		
20	(0,0;0),	(0,0;2)		
21	(0,0;1)			
22	(0,0;-1),	(0,0;3)		
23	(0,1;0),	(0,1;3),	(0,-1;-1),	(0,-1;2)
24	(0,1;1),	(0,1;2),	(0,-1;0),	(0,-1;1)
25	(0,2;1),	(0,2;3),	(0,-2;1),	(0,-2;-1)
26	(0,2;2),	(0,-2;0)		
27	(0,3;2),	(0,3;3),	(0,-3;0),	(0,-3;-1)
28	(1,-1;1)			
29	(1,-1;0),	(1,-1;2)		
30	(1,-2;0),	(1,-2;1),	(-1,2;1),	(-1,2;2)
31	(1,-3;0),	(-1,3;2)		
32	(1,1;1),	(1,1;3),	(-1,-1;1),	(-1,-1;-1)
33	(1,1;2),	(-1,-1;0)		
34	(1,2;2),	(1,2;3),	(-1,-2;0),	(-1,-2;-1)
35	(1,3;3),	(-1,-3;-1)		
36	(2,-2;1)			
37	(2,2;2),	(2,2;4),	(-2,-2;0),	(-2,-2;-2)
38	(2,2;3),	(-2,-2;-1)		
39	(2,3;3),	(2,3;4),	(-2,-3;-1),	(-2,-3;-2)
40	(2,4;4),	(-2,-4;-2)		
41	(3,3;4),	(-3,-3;-2)		
42	(3,4;4),	(3,4;5),	(-3,-4;-2),	(-3,-4;-3)
43	(4,4;5),	(-4,-4;-3)		

2 The q-Whipple Transformations

Recall the Watson (1929) transform (cf. [11, §2.5])

$$
{}_8\phi_7 \left[\begin{array}{cccccccc} a, & q\sqrt{a}, & -q\sqrt{a}, & b, & c, & d, & e, & q^{-n} \\ & \sqrt{a}, & -\sqrt{a}, & qa/b, & qa/c, & qa/d, & qa/e, & aq^{1+n} \end{array} ; \frac{q^{2+n}a^2}{bcde} \right] \quad (2.1a)
$$

$$
= \left[\begin{array}{cc} qa, & qa/bc \\ qa/b, & qa/c \end{array} ; q \right]_n \times {}_4\phi_3 \left[\begin{array}{cccc} q^{-n}, & b, & c, & qa/de \\ & qa/d, & qa/e, & q^{-n}bc/a \end{array} ; q \right] \quad (2.1b)
$$

which can be derived by the following series rearrangement

$$
[Eq:2.1a] = \sum_{m=0}^{n} \frac{1-aq^{2m}}{1-a} \left[\begin{array}{cccc} a, & b, & c, & q^{-n} \\ q, & qa/b, & qa/c, & q^{1+n}a \end{array} ; q \right]_m \left(\frac{q^{1+n}a}{bc}\right)^m
$$

$$
\times {}_3\phi_2 \left[\begin{array}{ccc} q^{-m}, & q^m a, & qa/de \\ & qa/d, & qa/e \end{array} ; q \right]
$$

$$
= \sum_{k=0}^{n} (-1)^k q^{nk-\binom{k}{2}} (qa/bc)^k \frac{(qa)_{2k}}{(q)_k} \left[\begin{array}{cccc} q^{-n}, & b, & c, & qa/de \\ q^{1+n}a, & qa/b, & qa/c, & qa/d, & qa/e \end{array} ; q \right]_k
$$

$$
\times {}_6\phi_5 \left[\begin{array}{cccccc} aq^{2k}, & q^{1+k}\sqrt{a}, & -q^{1+k}\sqrt{a}, & bq^k, & cq^k, & q^{k-n} \\ & q^k\sqrt{a}, & -q^k\sqrt{a}, & q^{1+k}a/b, & q^{1+k}a/c, & q^{1+n+k}a \end{array} ; \frac{q^{1+n-k}a}{bc} \right]
$$

$$
= [Eq:2.1b]
$$

in view of the the q-Saalschütz formula (1.1) and the q-Dixon formula (cf. [11, §2.4])

$$
{}_6\phi_5 \left[\begin{array}{cccccc} a, & q\sqrt{a}, & -q\sqrt{a}, & b, & c, & q^{-n} \\ & \sqrt{a}, & -\sqrt{a}, & qa/b, & qa/c, & q^{1+n}a \end{array} ; \frac{q^{1+n}a}{bc} \right] = \left[\begin{array}{cc} qa, & qa/bc \\ qa/b, & qa/c \end{array} ; q \right]_n .
$$

The symmetric properties of [Eq:2.1a] and [Eq:2.1b] on parameters b, c, d and e correspond to some particular balanced transformations specified from the Sears (1951) and Hall (1936) formulas (cf. [11, Appendix III])

$$
{}_4\phi_3 \left[\begin{array}{cccc} q^{-n}, & a, & b, & c \\ & d, & e, & f \end{array} ; q \right] \quad (2.2a)
$$

$$
= a^n \left[\begin{array}{cc} e/a, & f/a \\ e, & f \end{array} ; q \right]_n {}_4\phi_3 \left[\begin{array}{cccc} q^{-n}, & a, & d/b, & d/c \\ & d, & q^{1-n}a/e, & q^{1-n}a/f \end{array} ; q \right] \quad (2.2b)
$$

$$
= \left[\begin{array}{cc} a, & ef/ab, & ef/ac \\ e, & f, & ef/abc \end{array} ; q \right]_n {}_4\phi_3 \left[\begin{array}{cccc} q^{-n}, & ef/abc, & e/a, & f/a \\ & q^{1-n}/a, & ef/ab, & ef/ac \end{array} ; q \right] \quad (2.2c)
$$

where $q^{1-n}abc = def$.

Now, we perform the same manipulation on the basic almost poised series.

$$
\Omega_\delta \begin{bmatrix} \alpha, \gamma, \beta, \rho; \varepsilon \\ b, c, d, e; n \end{bmatrix} = {}_5\phi_4 \begin{bmatrix} q^{-\delta-2n}, & b, & c, & d, & e & ; \dfrac{q^{\varepsilon-3n}}{bcde} \\ & q^{\alpha-2n}/b, q^{\gamma-2n}/c, q^{\beta-2n}/d, q^{\rho-2n}/e \end{bmatrix} \quad (2.3\text{a})
$$

$$
= \sum_m \left(\frac{q^{\varepsilon+\lambda-\mu-\nu-n}}{bc} \right)^m {}_3\phi_2 \begin{bmatrix} q^{-m}, & q^{\lambda-1+m-2n}, & q^{\mu+\nu-\lambda-2n}/de \\ & q^{\mu-2n}/d, & q^{\nu-2n}/e \end{bmatrix} \quad (2.3\text{b})
$$

$$
\times \begin{bmatrix} q^{-\delta-2n}, & b, & c, & d, & e, & q^{\mu-2n}/d, \ q^{\nu-2n}/e \\ q, & q^{\alpha-2n}/b, & q^{\gamma-2n}/c, & dq^{\lambda-\mu}, & eq^{\lambda-\nu}, & q^{\beta-2n}/d, \ q^{\rho-2n}/e \end{bmatrix}_m \quad (2.3\text{c})
$$

$$
= \sum_k (-1)^k \frac{[q^{k+\lambda-1-2n}, q^{-\delta-2n}; q]_k}{(q; q)_k} q^{k(\varepsilon+\lambda-\mu-\nu-n)-\binom{k}{2}} / (bc)^k \quad (2.3\text{d})
$$

$$
\times \begin{bmatrix} b, & c, & d, & e, & q^{\mu+\nu-\lambda-2n}/de \\ q^{\alpha-2n}/b, & q^{\gamma-2n}/c, & dq^{\lambda-\mu}, & eq^{\lambda-\nu}, & q^{\beta-2n}/d, \ q^{\rho-2n}/e \end{bmatrix}_k \quad (2.3\text{e})
$$

$$
\times {}_8\phi_7 \begin{bmatrix} q^{\lambda-1-2n+2k}, bq^k, & cq^k, & dq^k, & eq^k, & q^{-\delta-2n+k}, \\ q^{\alpha-2n+k}/b, q^{\gamma-2n+k}/c, dq^{\lambda-\mu+k}, eq^{\lambda-\nu+k}, q^{\lambda-1-2n+k}, \end{bmatrix} \quad (2.3\text{f})
$$

$$
\begin{matrix} q^{\mu-2n+k}/d, & q^{\nu-2n+k}/e & ; \dfrac{q^{\varepsilon+\lambda-\mu-\nu-n-k}}{bc} \\ q^{\beta-2n+k}/d, & q^{\rho-2n+k}/e \end{matrix} \Biggr]^{(\delta+2n-k)} \quad (2.3\text{g})
$$

In this transformation, the additional parameters λ, μ and ν are assumed to be integral. They will be chosen so as to create the transformations of Whipple type.

By means of the decomposition

$$
1 - q^k w = (1 - q^k b) \frac{w-c}{b-c} + (1 - q^k c) \frac{b-w}{b-c}
$$

we may rewrite the ${}_4\phi_3$-series, when $deq^n = qbc$, in terms of two balanced series

$$
{}_4\phi_3 \begin{bmatrix} q^{-n}, & qw, & b, & c \\ & w, & d, & e \end{bmatrix} ; q \end{bmatrix} = \frac{1-b}{1-w} \frac{w-c}{b-c} {}_3\phi_2 \begin{bmatrix} q^{-n}, & bq, & c \\ & d, & e \end{bmatrix} ; q \end{bmatrix} \quad (2.4\text{a})
$$

$$
+ \frac{1-c}{1-w} \frac{b-w}{b-c} {}_3\phi_2 \begin{bmatrix} q^{-n}, & b, & cq \\ & d, & e \end{bmatrix} ; q \end{bmatrix} \quad (2.4\text{b})
$$

which will be used frequently to simplify the evaluation of the 2-balanced series.

2.1   Case $(\beta = \rho = 1 - \delta)$

Taking

$$\begin{aligned}
\beta = \rho &= 1 - \delta \\
\lambda = \mu = \nu &= 1 - \delta
\end{aligned}$$

in the general relation (2.3), the following transform may be established after some trivial modification.

Proposition I

$$\Omega_\delta \begin{bmatrix} \alpha, & \gamma, & 1-\delta, & 1-\delta; & \varepsilon \\ b, & c, & d, & e; & n \end{bmatrix} \tag{2.5a}$$

$$= \sum_k \frac{(q^{-\delta-2n}; q)_{2k}}{(q; q)_k} \begin{bmatrix} b, & c, & q^{1-\delta-2n}/de \\ q^{\alpha-2n}/b, & q^{\gamma-2n}/c, & q^{1-\delta-2n}/d, & q^{1-\delta-2n}/e \end{bmatrix}; q \Big]_k \tag{2.5b}$$

$$\times \frac{q^{k(\varepsilon+\delta-1-n)-\binom{k}{2}}}{(-bc)^k} \; {}_3\phi_2 \begin{bmatrix} q^{-\delta-2n+2k}, & bq^k, & cq^k & q^{\varepsilon+\delta-n-k-1} \\ & q^{\alpha-2n+k}/b, q^{\gamma-2n+k}/c; & bc \end{bmatrix} \tag{2.5c}$$

$$= \begin{bmatrix} q^{-\delta-2n}, & q^{[\alpha,\gamma]-2n}/bc; & q \\ q^{\alpha-2n}/b, & q^{\gamma-2n}/c \end{bmatrix}_n \sum_k \begin{bmatrix} q^{-\delta-n}, & b, & c, & q^{1-\delta-2n}/de; q \\ q, & q^{1-\delta-2n}/d, q^{1-\delta-2n}/e, bcq^{1+n-[\alpha,\gamma]} \end{bmatrix}_k \tag{2.5d}$$

$$\times \; q^{k(\varepsilon+\delta-[\alpha,\gamma])} \; \chi_\delta \begin{bmatrix} \alpha, & \gamma; & \varepsilon+\delta-1 \\ bq^k, & cq^k; & n-k \end{bmatrix} \tag{2.5e}$$

Replacing the χ-factor by the corresponding q-factorial product evaluations displayed in Section 1.2 and Section 1.3, we obtain the following transforms of the Whipple type.

2.1A   Transforms and the q-Dixon formulas when $\delta = 0$

Example 2.1 *For $(\alpha, \gamma) = (1, 1)$, the transforms of* Bailey [2] and Jackson [14] (see also [4, Eq.14]) *may be recovered*

$$\Omega_0 \begin{bmatrix} 1,1,1,1; & \varepsilon \\ b, c, d, e; & n \end{bmatrix} \xlongequal{\varepsilon=2,3} q^{(\varepsilon-2)n} \begin{bmatrix} q^{-2n}, & q^{1-2n}/bc \\ q^{1-2n}/b, & q^{1-2n}/c \end{bmatrix}; q \Big]_n$$

$$\times \; {}_4\phi_3 \begin{bmatrix} q^{-n}, & b, & c, & q^{1-2n}/de \\ & q^{1-2n}/d, q^{1-2n}/e, & bcq^n \end{bmatrix}; q \Big]$$

based on the evaluation from Section 1.2

$$\chi_0 \begin{bmatrix} 1, & 1; & \varepsilon-1 \\ bq^k, & cq^k; & n-k \end{bmatrix} = q^{(n-k)(\varepsilon-2)}.$$

When $bcde = q^{1-3n}$, *they reduce, by means of the Saalschütz formula* (1.1), *to the evaluations of* Jackson [14] (*cf.* [2], [4, Eq.1] *and* [16, Eq.3.8])

$$
{}_5\phi_4 \left[\begin{array}{ccccc} q^{-2n}, & b, & c, & d, & e \\ & q^{1-2n}/b, & q^{1-2n}/c, & q^{1-2n}/d & q^{1-2n}/e \end{array} ; q^\varepsilon \right]
$$

$$
\stackrel{\varepsilon=1,2}{=\!=\!=} \; q^{(\varepsilon-1)n} \left[\begin{array}{c} q^{-2n}, \; q^{1-2n}/bc, q^{1-2n}/bd, \; q^{1-2n}/cd \\ q^{1-2n}/b, \; q^{1-2n}/c, \; q^{1-2n}/d, \; q^{1-2n}/bcd \end{array} ; q \right]_n .
$$

Example 2.2 *For* $(\alpha, \gamma) = (0,2)$, *let*
$$
w = (1 + cq^{n-1})/(1 + bq^n).
$$

The new transforms read as

$$
\Omega_0 \left[\begin{array}{c} 0,2,1,1; \; \varepsilon \\ b, c, d, e; \; n \end{array} \right] \stackrel{\varepsilon=2,3}{=\!=\!=} q^{(\varepsilon-3)n} \left[\begin{array}{cccc} c, & bc/q, & q^{-2n}, & q^{2-2n}/bc \\ c/q, & bc, & q^{-2n}/b, & q^{2-2n}/c \end{array} ; q \right]_n
$$

$$
\times \left\{ \begin{array}{l} \left[\begin{array}{c} bw, \; -bq^n \\ bc/q \end{array} \right] {}_5\phi_4 \left[\begin{array}{cccc} q^{-n}, qbw & b, & c/q, & q^{1-2n}/de \\ & bw, q^{1-2n}/d, q^{1-2n}/e, bcq^n \end{array} ; q \right] \quad (\varepsilon = 2) \\[2em] \left[\begin{array}{c} q^{-1}c/w, -cq^{n-1} \\ bc/q \end{array} \right] {}_5\phi_4 \left[\begin{array}{cccc} q^{-n}, c/w & b, & c/q, & q^{1-2n}/de \\ q^{-1}c/w, q^{1-2n}/d, q^{1-2n}/e, & bcq^n \end{array} ; q \right] \quad (\varepsilon = 3) \end{array} \right.
$$

based on the reformulations from Section 1.2

$$
\chi_\delta \left[\begin{array}{ccc} 0, & 2; & \varepsilon - 1 \\ bq^k, & cq^k; & n-k \end{array} \right] \stackrel{\varepsilon=2,3}{=\!=\!=} \left\{ \begin{array}{l} \left[\begin{array}{c} cq^{n-1}, -bq^n \\ cq^{k-1}, bcq^{n+k-1} \end{array} \right] \times (1 - q^k bw) \, q^{k-n} \quad (\varepsilon = 2) \\[1.5em] \left[\begin{array}{c} cq^{n-1}, -cq^{n-1} \\ cq^{k-1}, bcq^{n+k-1} \end{array} \right] \times (1 - q^{k-1}c/w) \quad (\varepsilon = 3). \end{array} \right.
$$

When $bcde = q^{1-3n}$, *they reduce, by means of the 2-balanced separation* (2.4) *via substituting* $w \to bw$, $q^{-1}c/w$ *in accordance with* $\varepsilon = 2, 3$, *respectively, to evaluations*

$$
{}_5\phi_4 \left[\begin{array}{ccccc} q^{-2n}, & b, & c, & d, & e \\ & q^{-2n}/b, & q^{2-2n}/c, & q^{1-2n}/d, & q^{1-2n}/e \end{array} ; q^\varepsilon \right]
$$

$$
\stackrel{\varepsilon=1,2}{=\!=\!=} \left[\begin{array}{cccc} q^{-2n}, & q^{2-2n}/bc, & q^{-2n}/bd, & q^{2-2n}/cd \\ q^{-2n}/b, & q^{2-2n}/c, & q^{1-2n}/d, & q^{1-2n}/bcd \end{array} ; q \right]_n
$$

$$
\times \left\{ \begin{array}{l} \left[\begin{array}{c} b, q^{1-n}/c \\ q/c, bcq^{n-1} \end{array} \right] \left\{ 1 + bq^n \left[\begin{array}{c} c/q, bdq^n, cdq^{2n-1} \\ b, bd^{2n}, cdq^{n-1} \end{array} \right] \right\} \quad (\varepsilon = 1) \\[2em] \left[\begin{array}{c} 1/b, cq^{n-1} \\ c/q, q^{1-n}/bc \end{array} \right] \left\{ 1 + q^{1-n}/c \left[\begin{array}{c} c/q, bdq^n, cdq^{2n-1} \\ b, bd^{2n}, cdq^{n-1} \end{array} \right] \right\} \quad (\varepsilon = 2). \end{array} \right.
$$

Example 2.3 *For* $(\alpha, \gamma) = (1,2)$, *the new transform reads as*

$$
\Omega_0 \left[\begin{array}{c} 1,2,1,1; \; 3 \\ b, c, d, e; \; n \end{array} \right] = \left[\begin{array}{cccc} c, & q^{-2n}, & q^{2-2n}/bc \\ c/q, & q^{1-2n}/b, & q^{2-2n}/c \end{array} ; q \right]_n
$$

$$
\times {}_4\phi_3 \left[\begin{array}{cccc} q^{-n}, & b, & c/q, & q^{1-2n}/de \\ & q^{1-2n}/d, q^{1-2n}/e, & bcq^{n-1} \end{array} ; q \right]
$$

based on the evaluation from Section 1.2

$$\chi_0 \begin{bmatrix} 1, & 2; & 2 \\ bq^k, & cq^k; & n-k \end{bmatrix} = \frac{1 - cq^{n-1}}{1 - cq^{k-1}}.$$

When $bcde = q^{2-3n}$, it reduces, by means of the Saalschütz formula (1.1), to the evaluation of Jain and Verma [15, Eq.5.1]

$$_5\phi_4 \begin{bmatrix} q^{-2n}, & b, & c, & d, & e \\ & q^{1-2n}/b, & q^{2-2n}/c, & q^{1-2n}/d, & q^{1-2n}/e \end{bmatrix}; q \end{bmatrix}$$

$$= \begin{bmatrix} c, & q^{-2n}, & q^{2-2n}/bc, & q^{1-2n}/bd, & q^{2-2n}/cd \\ c/q, & q^{1-2n}/b, & q^{2-2n}/c, & q^{1-2n}/d, & q^{2-2n}/bcd \end{bmatrix}; q \end{bmatrix}_n.$$

Example 2.4 *For $(\alpha, \gamma) = (0, 1)$, the new transform reads as*

$$\Omega_0 \begin{bmatrix} 0, 1, 1, 1; & 2 \\ b, c, d, e; & n \end{bmatrix} = \begin{bmatrix} q^{-2n}, & q^{1-2n}/bc \\ q^{-2n}/b, & q^{1-2n}/c \end{bmatrix}; q \end{bmatrix}_n$$

$$\times \quad _4\phi_3 \begin{bmatrix} q^{-n}, & b, & c, & q^{1-2n}/de \\ & q^{1-2n}/d, & q^{1-2n}/e, & bcq^n \end{bmatrix}; q \end{bmatrix}$$

based on the evaluation from Section 1.2

$$\chi_0 \begin{bmatrix} 0, & 1; & 1 \\ bq^k, & cq^k; & n-k \end{bmatrix} = 1.$$

When $bcde = q^{1-3n}$, it reduces, by means of the Saalschütz formula (1.1), to

$$_5\phi_4 \begin{bmatrix} q^{-2n}, & b, & c, & d, & e \\ & q^{-2n}/b, & q^{1-2n}/c, & q^{1-2n}/d, & q^{1-2n}/e \end{bmatrix}; q \end{bmatrix}$$

$$= \begin{bmatrix} q^{-2n}, & q^{1-2n}/bc, & q^{1-2n}/bd, & q^{1-2n}/cd \\ q^{-2n}/b, & q^{1-2n}/c, & q^{1-2n}/d, & q^{1-2n}/bcd \end{bmatrix}; q \end{bmatrix}_n.$$

This is the reversal version of Example 2.3.

Example 2.5 *For $(\alpha, \gamma) = (2, 2)$, the new transforms read as*

$$\Omega_0 \begin{bmatrix} 2, 2, 1, 1; & \varepsilon \\ b, c, d, e; & n \end{bmatrix} \xlongequal{\varepsilon=3,4} q^{n(2\varepsilon-7)} \begin{bmatrix} b, & c, & q^{-2n}, & q^{2-2n}/bc \\ b/q, & c/q, & q^{2-2n}/b, & q^{2-2n}/c \end{bmatrix}; q \end{bmatrix}_n$$

$$\times \quad _4\phi_3 \begin{bmatrix} q^{-n}, & b/q, & c/q, & q^{1-2n}/de \\ & q^{1-2n}/d, q^{1-2n}/e, & bcq^{n-1} \end{bmatrix}; q^{5-\varepsilon} \end{bmatrix}$$

based on the evaluation from Section 1.2

$$\chi_0 \begin{bmatrix} 2, & 2; & \varepsilon - 1 \\ bq^k, & cq^k; & n-k \end{bmatrix} = q^{(n-k)(2\varepsilon-7)} \frac{(1 - bq^{n-1})(1 - cq^{n-1})}{(1 - bq^{k-1})(1 - cq^{k-1})}.$$

When $bcde = q^{2-3n}$, they reduce, by means of the 2-balanced separation (2.4) via substituting $w \to \infty, 0$ in accordance with $\varepsilon = 3, 4$, respectively, to

$$_5\phi_4 \begin{bmatrix} q^{-2n}, & b, & c, & d, & e \\ & q^{2-2n}/b, & q^{2-2n}/c, & q^{1-2n}/d, & q^{1-2n}/e \end{bmatrix}; q^\varepsilon \end{bmatrix}$$

$$\xlongequal{\varepsilon=1,2} q^{n(\varepsilon-2)} \left\{ 1 - (q^{2n-2}bcd)^{\varepsilon-1} \begin{bmatrix} q^{-n}, & dq^n \\ bdq^{n-1}, & cdq^{n-1} \end{bmatrix} \right\}$$

$$\times \begin{bmatrix} b, c, & q^{-2n}, & q^{2-2n}/bc, & q^{2-2n}/bd, & q^{2-2n}/cd \\ b/q, c/q, & q^{2-2n}/b, & q^{2-2n}/c, & q^{1-2n}/d, & q^{2-2n}/bcd \end{bmatrix}; q \end{bmatrix}_n.$$

Example 2.6 *For* $(\alpha, \gamma) = (0,0)$, *the transforms of* Bressoud *[4, Eqs.3,16]* *may be recovered*

$$\Omega_0 \begin{bmatrix} 0,0,1,1; & \varepsilon \\ b,c,d,e; & n \end{bmatrix} \overset{\varepsilon=1,2}{=\!=\!=} \begin{bmatrix} q^{-2n}, & q^{-2n}/bc \\ q^{-2n}/b, & q^{-2n}/c \end{bmatrix} ; q \Big]_n$$

$$\times \ {}_4\phi_3 \begin{bmatrix} q^{-n}, & b, & c, & q^{1-2n}/de \\ & q^{1-2n}/d, & q^{1-2n}/e, & bcq^{n+1} \end{bmatrix} ; q^\varepsilon \Big]$$

based on the evaluation from Section 1.2

$$\chi_0 \begin{bmatrix} 0, & 0; & \varepsilon - 1 \\ bq^k, & cq^k; & n-k \end{bmatrix} \ = \ 1.$$

When $bcde = q^{-3n}$, *they reduce, by means of the 2-balanced separation (2.4) via substituting* $w \to 0, \infty$ *in accordance with* $\varepsilon = 1, 2$, *respectively, to*

$$\ {}_5\phi_4 \begin{bmatrix} q^{-2n}, & b, & c, & d, & e \\ & q^{-2n}/b, & q^{-2n}/c, & q^{1-2n}/d, & q^{1-2n}/e \end{bmatrix} ; q^\varepsilon \Big]$$

$$\overset{\varepsilon=1,2}{=\!=\!=} \qquad q^{n(\varepsilon-2)} \quad \left\{ 1 - (q^{2n}bcd)^{2-\varepsilon} \begin{bmatrix} q^{-n}, & dq^n \\ bdq^n, & cdq^n \end{bmatrix} \right\}$$

$$\times \begin{bmatrix} q^{-2n}, & q^{-2n}/bc, & q^{1-2n}/bd, & q^{1-2n}/cd \\ q^{-2n}/b, & q^{-2n}/c, & q^{1-2n}/d, & q^{-2n}/bcd \end{bmatrix} ; q \Big]_n .$$

This is the reversal version of Example 2.5.

Example 2.7 *For* $(\alpha, \gamma) = (2,3)$, *the new transform reads as*

$$\Omega_0 \begin{bmatrix} 2,3,1,1; & 4 \\ b,c,d,e; & n \end{bmatrix} = q^{-3n} \begin{bmatrix} b, & c \\ b/q, & c/q^2 \end{bmatrix} ; q \Big]_{2n} \times \begin{bmatrix} q^{-2n}, & q^{3-2n}/bc \\ q^{1-2n}/b, & q^{2-2n}/c \end{bmatrix} ; q \Big]_n$$

$$\times \ {}_5\phi_4 \begin{bmatrix} q^{-n}, q^{2-2n}/c, & b/q, & c/q^2, & q^{1-2n}/de \\ q^{1-2n}/c, q^{1-2n}/d, q^{1-2n}/e, & bcq^{n-2} \end{bmatrix} ; q \Big]$$

based on the evaluation from Section 1.2

$$\chi_0 \begin{bmatrix} 2, & 3; & 3 \\ bq^k, & cq^k; & n-k \end{bmatrix} \ = \ q^{k-n} \frac{(1 - bq^{n-1})(1 - cq^{2n-k-1})(1 - cq^{n-2})}{(1 - bq^{k-1})(1 - cq^{k-1})(1 - cq^{k-2})}.$$

When $bcde = q^{3-3n}$, *it reduces, by means of the 2-balanced separation (2.4) via substituting* $w \to q^{1-2n}/c$, *to*

$$\ {}_5\phi_4 \begin{bmatrix} q^{-2n}, & b, & c, & d, & e \\ & q^{2-2n}/b, & q^{3-2n}/c, & q^{1-2n}/d, & q^{1-2n}/e \end{bmatrix} ; q \Big]$$

$$= q^{-n} \begin{bmatrix} b, c \\ b/q, c/q^2 \end{bmatrix} ; q \Big]_n \left\{ 1 - \begin{bmatrix} q^{-n}, & dq^n, & bc^2dq^{3n-4} \\ cq^{n-1}, & bdq^{n-1}, & cdq^{n-2} \end{bmatrix} \right\}$$

$$\times \begin{bmatrix} q^{-2n}, & q^{3-2n}/bc, & q^{2-2n}/bd, & q^{3-2n}/cd \\ q^{2-2n}/b, & q^{3-2n}/c, & q^{1-2n}/d, & q^{3-2n}/bcd \end{bmatrix} ; q \Big]_n .$$

Example 2.8 *For* $(\alpha, \gamma) = (0, -1)$, *the new transform reads as*

$$\Omega_0 \begin{bmatrix} 0, -1, 1, 1; & 1 \\ b, & c, & d, e; & n \end{bmatrix} = \begin{bmatrix} q^{-2n}, & q^{-2n}/bc \\ q^{-2n}/b, & q^{-2n}/c \end{bmatrix}; q \end{bmatrix}_n$$

$$\times \; {}_5\phi_4 \begin{bmatrix} q^{-n}, q^{-2n}/c, & b, & c, & q^{1-2n}/de \\ q^{-1-2n}/c, q^{1-2n}/d, q^{1-2n}/e, & bcq^{n+1} \end{bmatrix}; q \end{bmatrix}$$

based on the evaluation from Section 1.2

$$\chi_0 \begin{bmatrix} 0, & -1; & 0 \\ bq^k, & cq^k; & n-k \end{bmatrix} = q^{k-n} \frac{1 - cq^{1+2n-k}}{1 - cq^{1+n}}.$$

When $bcde = q^{-3n}$, *it reduces, by means of the 2-balanced separation* (2.4) *via substituting* $w \to q^{-1-2n}/c$, *to*

$$_5\phi_4 \begin{bmatrix} q^{-2n}, & b, & c, & d, & e \\ & q^{-2n}/b, & q^{-1-2n}/c, & q^{1-2n}/d, & q^{1-2n}/e \end{bmatrix}; q \end{bmatrix}$$

$$= \quad q^{-n} \left\{ 1 - \begin{bmatrix} q^{-n}, & dq^n, & bc^2dq^{3n+1} \\ cq^{n+1}, & bdq^n, & cdq^n \end{bmatrix} \right\}$$

$$\times \begin{bmatrix} q^{-2n}, & q^{-2n}/bc, & q^{1-2n}/bd, & q^{1-2n}/cd \\ q^{-2n}/b, & q^{-1-2n}/c, & q^{1-2n}/d, & q^{-2n}/bcd \end{bmatrix}; q \end{bmatrix}_n.$$

This is the reversal version of Example 2.7.

2.1B   Transforms and the q-Dixon formulas when $\delta = 1$

Example 2.9 *For* $(\alpha, \gamma) = (0, 0)$, *the transforms read as*

$$\Omega_1 \begin{bmatrix} 0, 0, 0, 0; & \varepsilon \\ b, & c, & d, e; & n \end{bmatrix} \xlongequal{\varepsilon = 0, 1, 2} (1 - q^{(\varepsilon - 1)(2n+1)}) \begin{bmatrix} q^{-2n}, & q^{-2n}/bc \\ q^{-2n}/b, & q^{-2n}/c \end{bmatrix}; q \end{bmatrix}_n$$

$$\times \; {}_4\phi_3 \begin{bmatrix} q^{-n}, & b, & c, & q^{-2n}/de \\ & q^{-2n}/d, & q^{-2n}/e, & bcq^{n+1} \end{bmatrix}; q \end{bmatrix}$$

based on the evaluation from Section 1.3

$$\chi_1 \begin{bmatrix} 0, & 0; & \varepsilon \\ bq^k, & cq^k; & n-k \end{bmatrix} = \begin{cases} 1 - q^{k-n-1}, & \varepsilon = 0 \\ 0, & \varepsilon = 1 \\ q^{n-k}(1 - q^{n-k+1}), & \varepsilon = 2. \end{cases}$$

Especially for $\varepsilon = 1$, *the corresponding identity*

$$_5\phi_4 \begin{bmatrix} q^{-1-2n}, & b, & c, & d, & e \\ & q^{-2n}/b, & q^{-2n}/c, & q^{-2n}/d, & q^{-2n}/e \end{bmatrix}; \frac{q^{1-3n}}{bcde} \end{bmatrix} = 0$$

is an extension of [16, Eq.3.18] *due to* Joshi *and* Verma. *When* $bcde = q^{-1-3n}$, *they reduce, by means of the Saalschütz theorem* (1.1), *to the evaluations of* Carlitz [5, Eq.1.7 & Eq.3.5] *and* Joshi & Verma [16, Eq.3.12]

$$_5\phi_4 \begin{bmatrix} q^{-1-2n}, & b, & c, & d, & e \\ & q^{-2n}/b, & q^{-2n}/c, & q^{-2n}/d, & q^{-2n}/e \end{bmatrix}; q^\varepsilon \end{bmatrix}$$

$$\xlongequal{\varepsilon = 1, 2, 3} (1 - q^{(\varepsilon - 2)(2n+1)}) \begin{bmatrix} q^{-2n}, & q^{-2n}/bc, & q^{-2n}/bd, & q^{-2n}/cd \\ q^{-2n}/b, & q^{-2n}/c, & q^{-2n}/d, & q^{-2n}/bcd \end{bmatrix}; q \end{bmatrix}_n.$$

Example 2.10 *For $(\alpha, \gamma) = (1, -1)$, the new transform reads as*

$$\Omega_1 \begin{bmatrix} 1, -1, 0, 0; \ 1 \\ b, \ c, \ d, e; \ n \end{bmatrix} = \begin{bmatrix} qc/b, & q^{2n+1} \\ q/b, & cq^{2n+1} \end{bmatrix} \times \begin{bmatrix} q^{-2n}, & q^{1-2n}/bc \\ q^{1-2n}/b, & q^{-2n}/c \end{bmatrix}; q \end{bmatrix}_n$$

$$\times \ {}_4\phi_3 \begin{bmatrix} q^{-n}, & b/q, & c, & q^{-2n}/de \\ & q^{-2n}/d, & q^{-2n}/e, & bcq^n \end{bmatrix}; q \end{bmatrix}$$

based on the evaluation from Section 1.3

$$\chi_1 \begin{bmatrix} 1, & -1; & 1 \\ bq^k, & cq^k; & n-k \end{bmatrix} = \frac{(1 - q^{1+n-k})(1 - qc/b)}{(1 - cq^{1+n})(1 - q^{1-k}/b)}.$$

When $bcde = q^{-3n}$, it reduces, by means of the Saalschütz theorem (1.1), to

$${}_5\phi_4 \begin{bmatrix} q^{-1-2n}, & b, & c, & d, & e \\ & q^{1-2n}/b, & q^{-1-2n}/c, & q^{-2n}/d, & q^{-2n}/e \end{bmatrix}; q \end{bmatrix}$$

$$= \begin{bmatrix} qc/b, & q^{2n+1} \\ q/b, & cq^{2n+1} \end{bmatrix} \times \begin{bmatrix} q^{-2n}, & q^{1-2n}/bc, & q^{1-2n}/bd, & q^{-2n}/cd \\ q^{1-2n}/b, & q^{-2n}/c, & q^{-2n}/d, & q^{1-2n}/bcd \end{bmatrix}; q \end{bmatrix}_n.$$

Example 2.11 *For $(\alpha, \gamma) = (0, 1)$, the new transforms read as*

$$\Omega_1 \begin{bmatrix} 0, 1, 0, 0; \ \varepsilon \\ b, c, d, e; \ n \end{bmatrix} \xlongequal{\varepsilon = 1, 2} (cq^{2n})^{\varepsilon - 2} \frac{1 - q^{2n+1}}{1 - q/c} \begin{bmatrix} q^{-2n}, & q^{1-2n}/bc \\ q^{-2n}/b, & q^{1-2n}/c \end{bmatrix}; q \end{bmatrix}_n$$

$$\times \ {}_4\phi_3 \begin{bmatrix} q^{-n}, & b, & c/q, & q^{-2n}/de \\ & q^{-2n}/d, & q^{-2n}/e, & bcq^n \end{bmatrix}; q \end{bmatrix}$$

based on the evaluation from Section 1.3

$$\chi_1 \begin{bmatrix} 0, & 1; & \varepsilon \\ bq^k, & cq^k; & n-k \end{bmatrix} = (cq^{2n-k})^{\varepsilon - 1} \frac{1 - q^{k-n-1}}{1 - cq^{k-1}}.$$

When $bcde = q^{-3n}$, they reduce, by means of the Saalschütz theorem (1.1), to

$${}_5\phi_4 \begin{bmatrix} q^{-1-2n}, & b, & c, & d, & e \\ & q^{-2n}/b, & q^{1-2n}/c, & q^{-2n}/d, & q^{-2n}/e \end{bmatrix}; q^\varepsilon \end{bmatrix}$$

$$\xlongequal{\varepsilon = 1, 2} (cq^{2n})^{\varepsilon - 2} \frac{1 - q^{2n+1}}{1 - q/c} \begin{bmatrix} q^{-2n}, & q^{1-2n}/bc, & q^{-2n}/bd, & q^{1-2n}/cd \\ q^{-2n}/b, & q^{1-2n}/c, & q^{-2n}/d, & q^{1-2n}/bcd \end{bmatrix}; q \end{bmatrix}_n.$$

Example 2.12 *For $(\alpha, \gamma) = (0, -1)$, the new transforms read as*

$$\Omega_1 \begin{bmatrix} 0, -1, 0, 0; \ \varepsilon \\ b, \ c, \ d, e; \ n \end{bmatrix} \xlongequal{\varepsilon = 0, 1} c^{1-\varepsilon} \frac{1 - q^{2n+1}}{1 - cq^{2n+1}} \begin{bmatrix} q^{-2n}, & q^{-2n}/bc \\ q^{-2n}/b, & q^{-2n}/c \end{bmatrix}; q \end{bmatrix}_n$$

$$\times \ {}_4\phi_3 \begin{bmatrix} q^{-n}, & b, & c, & q^{-2n}/de \\ & q^{-2n}/d, & q^{-2n}/e, & bcq^{n+1} \end{bmatrix}; q \end{bmatrix}$$

based on the evaluation from Section 1.3

$$\chi_1 \begin{bmatrix} 0, & -1; & \varepsilon \\ bq^k, & cq^k; & n-k \end{bmatrix} = c^{1-\varepsilon} \frac{1 - q^{1+n-k}}{1 - cq^{1+n}}.$$

When $bcde = q^{-1-3n}$, they reduce, by means of the Saalschütz theorem (1.1), to

$$
{}_5\phi_4\left[\begin{array}{ccccc} q^{-1-2n}, & b, & c, & d, & e \\ & q^{-2n}/b, & q^{-1-2n}/c, & q^{-2n}/d, & q^{-2n}/e \end{array}; q^\varepsilon\right]
$$

$$
\underset{\varepsilon=1,2}{=\!=\!=} \quad c^{2-\varepsilon}\,\frac{1-q^{2n+1}}{1-cq^{2n+1}}\left[\begin{array}{c} q^{-2n},\, q^{-2n}/bc,\, q^{-2n}/bd,\, q^{-2n}/cd \\ q^{-2n}/b,\, q^{-2n}/c,\, q^{-2n}/d,\, q^{-2n}/bcd \end{array}; q\right]_n .
$$

This is the reversal version of Example 2.11.

Example 2.13 *For* $(\alpha,\gamma) = (1,1)$*, the new transform reads as*

$$
\Omega_1\left[\begin{array}{c} 1,1,0,0;\ 2 \\ b,c,d,e;\ n \end{array}\right] = \left[\begin{array}{c} q^{-1-2n},\, bcq^{2n-1} \\ b/q,\, c/q \end{array}\right] \times \left[\begin{array}{c} q^{-2n},\, q^{2-2n}/bc \\ q^{1-2n}/b,\, q^{1-2n}/c \end{array}; q\right]_n
$$

$$
\times\ {}_4\phi_3\left[\begin{array}{cccc} q^{-n}, & b/q, & c/q, & q^{-2n}/de \\ & q^{-2n}/d, & q^{-2n}/e, & bcq^{n-1} \end{array}; q\right]
$$

based on the evaluation from Section 1.3

$$
\chi_1\left[\begin{array}{ccc} 1, & 1; & 2 \\ bq^k, & cq^k; & n-k \end{array}\right] = q^{n-k}\,\frac{(1-q^{k-n-1})\,(1-bcq^{k+n-1})}{(1-bq^{k-1})\,(1-cq^{k-1})} .
$$

When $bcde = q^{1-3n}$*, it reduces, by means of the Saalschütz theorem (1.1), to*

$$
{}_5\phi_4\left[\begin{array}{ccccc} q^{-1-2n}, & b, & c, & d, & e \\ & q^{1-2n}/b, & q^{1-2n}/c, & q^{-2n}/d, & q^{-2n}/e \end{array}; q\right]
$$

$$
= \left[\begin{array}{c} q^{-1-2n},\, bcq^{2n-1} \\ b/q,\, c/q \end{array}\right] \times \left[\begin{array}{c} q^{-2n},\, q^{2-2n}/bc,\, q^{1-2n}/bd,\, q^{1-2n}/cd \\ q^{1-2n}/b,\, q^{1-2n}/c,\, q^{-2n}/d,\, q^{2-2n}/bcd \end{array}; q\right]_n .
$$

Example 2.14 *For* $(\alpha,\gamma) = (-1,-1)$*, the new transform reads as*

$$
\Omega_1\left[\begin{array}{c} -1,-1,0,0;\ 0 \\ b,\ c,\ d,e;\ n \end{array}\right] = \left[\begin{array}{c} q^{-1-2n}, \quad q^{-1-2n}/bc \\ q^{-1-2n}/b, \quad q^{-1-2n}/c \end{array}; q\right]_{n+1}
$$

$$
\times\ {}_4\phi_3\left[\begin{array}{cccc} q^{-n}, & b, & c, & q^{-2n}/de \\ & q^{-2n}/d, & q^{-2n}/e, & bcq^{n+1} \end{array}; q\right]
$$

based on the evaluation from Section 1.3

$$
\chi_1\left[\begin{array}{ccc} -1, & -1; & 0 \\ bq^k, & cq^k; & n-k \end{array}\right] = \frac{(1-q^{1+n-k})\,(1-bcq^{1+n+k})}{(1-bq^{1+n})\,(1-cq^{1+n})} .
$$

When $bcde = q^{-1-3n}$*, it reduces, by means of the Saalschütz theorem (1.1), to*

$$
{}_5\phi_4\left[\begin{array}{ccccc} q^{-1-2n}, & b, & c, & d, & e \\ & q^{-1-2n}/b, & q^{-1-2n}/c, & q^{-2n}/d, & q^{-2n}/e \end{array}; q\right]
$$

$$
= \left[\begin{array}{c} q^{2n+1}, \quad bcq^{2n+1} \\ bq^{2n+1}, \quad cq^{2n+1} \end{array}\right] \times \left[\begin{array}{c} q^{-2n},\, q^{-2n}/bc,\, q^{-2n}/bd,\, q^{-2n}/cd \\ q^{-2n}/b,\, q^{-2n}/c,\, q^{-2n}/d,\, q^{-2n}/bcd \end{array}; q\right]_n .
$$

This is the reversal version of Example 2.13.

Example 2.15 *For* $(\alpha, \gamma) = (0, 2)$, *the new transform reads as*

$$\Omega_1 \begin{bmatrix} 0,2,0,0; \ 2 \\ b, c, d, e; \ n \end{bmatrix} = \begin{bmatrix} q^{-1-2n}, \ c^2 q^{2n-2} \\ c/q, \ c/q^2 \end{bmatrix} \times \begin{bmatrix} q^{-2n}, \ q^{2-2n}/bc \\ q^{-2n}/b, \ q^{2-2n}/c \end{bmatrix}_n$$

$$\times \ {}_4\phi_3 \begin{bmatrix} q^{-n}, & b, & c/q^2, & q^{-2n}/de \\ & q^{-2n}/d, & q^{-2n}/e, & bcq^{n-1} \end{bmatrix} ; q \end{bmatrix}$$

based on the evaluation from Section 1.3

$$\chi_1 \begin{bmatrix} 0, & 2; & 2 \\ bq^k, & cq^k; & n-k \end{bmatrix} = \frac{(1 - q^{k-n-1})(1 - c^2 q^{2n-2})}{(1 - cq^{k-1})(1 - cq^{k-2})}.$$

When $bcde = q^{1-3n}$, *it reduces, by means of the Saalschütz theorem* (1.1), *to*

$${}_5\phi_4 \begin{bmatrix} q^{-1-2n}, & b, & c, & d, & e \\ & q^{-2n}/b, & q^{2-2n}/c, & q^{-2n}/d, & q^{-2n}/e \end{bmatrix} ; q \end{bmatrix}$$

$$= \begin{bmatrix} q^{-1-2n}, \ c^2 q^{2n-2} \\ c/q, \ c/q^2 \end{bmatrix} \times \begin{bmatrix} q^{-2n}, \ q^{2-2n}/bc, \ q^{-2n}/bd, \ q^{2-2n}/cd \\ q^{-2n}/b, \ q^{2-2n}/c, \ q^{-2n}/d, \ q^{2-2n}/bcd \end{bmatrix}_n.$$

Example 2.16 *For* $(\alpha, \gamma) = (0, -2)$, *the new transform reads as*

$$\Omega_1 \begin{bmatrix} 0,-2,0,0; \ 0 \\ b, \ c, \ d, e; \ n \end{bmatrix} = \begin{bmatrix} bq^n, \ -cq^{n+1} \\ bcq^{2n+1} \end{bmatrix} \times \begin{bmatrix} q^{-1-2n}, \ q^{-1-2n}/bc \\ q^{-2n}/b, \ q^{-2-2n}/c \end{bmatrix}_{n+1}$$

$$\times \ {}_4\phi_3 \begin{bmatrix} q^{-n}, & b, & c, & q^{-2n}/de \\ & q^{-2n}/d, & q^{-2n}/e, & bcq^{n+1} \end{bmatrix} ; q \end{bmatrix}$$

based on the evaluation from Section 1.3

$$\chi_1 \begin{bmatrix} 0, & -2; & 0 \\ bq^k, & cq^k; & n-k \end{bmatrix} = q^{k-n} \frac{(1 - q^{1+n-k})(1 + cq^{1+n})}{1 - cq^{2+n}}.$$

When $bcde = q^{-1-3n}$, *it reduces, by means of the Saalschütz theorem* (1.1), *to*

$${}_5\phi_4 \begin{bmatrix} q^{-1-2n}, & b, & c, & d, & e \\ & q^{-2n}/b, & q^{-2-2n}/c, & q^{-2n}/d, & q^{-2n}/e \end{bmatrix} ; q \end{bmatrix}$$

$$= q^{-2n} \begin{bmatrix} q^{2n+1}, \ -cq^{n+1} \\ cq^{n+2} \end{bmatrix} \times \begin{bmatrix} q^{-2n}, \ q^{-2n}/bc, \ q^{-2n}/bd, \ q^{-2n}/cd \\ q^{-2n}/b, \ q^{-2-2n}/c, \ q^{-2n}/d, \ q^{-2n}/bcd \end{bmatrix}_n.$$

This is the reversal version of **Example 2.15**.

Example 2.17 *For* $(\alpha, \gamma) = (2, 2)$, *the new transform reads as*

$$\Omega_1 \begin{bmatrix} 2,2,0,0; \ 3 \\ b, c, d, e; \ n \end{bmatrix} = -q^2 \begin{bmatrix} b, \ c \\ b/q^2, \ c/q^2 \end{bmatrix}_n \times \begin{bmatrix} q^{-2-2n}, \ q^{2-2n}/bc \\ q^{2-2n}/b, \ q^{2-2n}/c \end{bmatrix}_{n+1}$$

$$\times \ {}_5\phi_4 \begin{bmatrix} q^{-n}, -q^{-n} & b/q^2, & c/q^2, q^{-2n}/de \\ -q^{-n-1}, q^{-2n}/d, & q^{-2n}/e, & bcq^{n-2} \end{bmatrix} ; q \end{bmatrix}$$

based on the evaluation from Section 1.3

$$\chi_1 \begin{bmatrix} 2, & 2; & 3 \\ bq^k, & cq^k; & n-k \end{bmatrix} = q^{n-k} \frac{(1 - bq^{n-1})(1 - cq^{n-1})}{(1 - bq^{k-1})(1 - cq^{k-1})}$$

$$\times \frac{(1 - q^{2k-2n-2})(1 - bcq^{k+n-2})}{(1 - bq^{k-2})(1 - cq^{k-2})}.$$

When $bcde = q^{2-3n}$, it reduces, by means of the 2-balanced separation (2.4) via substituting $w \to -q^{-n-1}$, to

$$
{}_5\phi_4 \left[\begin{array}{ccccc} q^{-1-2n}, & b, & c, & d, & e \\ & q^{2-2n}/b, & q^{2-2n}/c, & q^{-2n}/d, & q^{-2n}/e \end{array} ; q \right]
$$
$$
= \left\{ 1 - \left[\begin{array}{c} q^{-n}, dq^{n+1}, -bcdq^{2n-2} \\ -q, \; bdq^{n-1}, \; cdq^{n-1} \end{array} \right] \right\} \times \left[\begin{array}{cccc} q^{-1-n}, bq^{n-1}, cq^{n-1}, bcq^{n-2} \\ b/q, \; b/q^2, \; c/q, \; c/q^2 \end{array} \right]
$$
$$
\times \quad (1 + q^{-1}) \; \left[\begin{array}{cccc} q^{-1-2n}, & q^{2-2n}/bc, & q^{2-2n}/bd, & q^{2-2n}/cd \\ q^{2-2n}/b, & q^{2-2n}/c, & q^{-2n}/d, & q^{3-2n}/bcd \end{array} ; q \right]_n .
$$

Example 2.18 *For* $(\alpha, \gamma) = (-2, -2)$, *the new transform reads as*

$$
\Omega_1 \left[\begin{array}{cccc} -2, & -2, 0, 0; & -1 \\ b, & c, \; d, e; & n \end{array} \right] = \left[\begin{array}{cc} q^{-2-2n}, & q^{-2-2n}/bc \\ q^{-2-2n}/b, & q^{-2-2n}/c \end{array} ; q \right]_{n+1}
$$
$$
\times \quad {}_5\phi_4 \left[\begin{array}{ccccc} q^{-n}, & -q^{-n}, & b, & c, & q^{-2n}/de \\ & -q^{-n-1}, & q^{-2n}/d, & q^{-2n}/e, & bcq^{n+2} \end{array} ; q \right]
$$

based on the evaluation from Section 1.3

$$
\chi_1 \left[\begin{array}{cc} -2, & -2; \; -1 \\ bq^k, & cq^k; \; n-k \end{array} \right] = q^{k-n} \frac{(1 - q^{2+2n-2k})(1 - bcq^{2+n+k})}{(1 - bq^{2+n})(1 - cq^{2+n})} .
$$

When $bcde = q^{-2-3n}$, it reduces, by means of the 2-balanced separation (2.4) via substituting $w \to -q^{-n-1}$, to

$$
{}_5\phi_4 \left[\begin{array}{ccccc} q^{-1-2n}, & b, & c, & d, & e \\ & q^{-2-2n}/b, & q^{-2-2n}/c, & q^{-2n}/d, & q^{-2n}/e \end{array} ; q \right]
$$
$$
= \left\{ 1 - \left[\begin{array}{c} q^{-n}, dq^{n+1}, -bcdq^{2n+2} \\ -q, \; bdq^{n+1}, \; cdq^{n+1} \end{array} \right] \right\} \times \left[\begin{array}{cc} q^{n+1}, & bcq^{2n+2} \\ bq^{2n+2}, & cq^{2n+2} \end{array} \right]
$$
$$
\times \quad (1 + q) \; \left[\begin{array}{cccc} q^{-1-2n}, & q^{-1-2n}/bc, & q^{-2n}/bd, & q^{-2n}/cd \\ q^{-1-2n}/b, & q^{-1-2n}/c, & q^{-2n}/d, & q^{-1-2n}/bcd \end{array} ; q \right]_n .
$$

This is the reversal version of Example 2.17.

2.2   Case ($\beta = \rho = -\delta$)

Taking

$$\beta = \rho \;=\; -\delta$$
$$\lambda = \mu = \nu \;=\; -\delta$$

in the general relation (2.3), the following transform may be established after some trivial modification.

Proposition II

$$\Omega_\delta \begin{bmatrix} \alpha, & \gamma, & -\delta, & -\delta; & \varepsilon \\ b, & c, & d, & e; & n \end{bmatrix} \tag{2.6a}$$

$$= \sum_k \begin{bmatrix} b, & c, & q^{-\delta-2n}/de \\ q^{\alpha-2n}/b, & q^{\gamma-2n}/c, & q^{-\delta-2n}/d, & q^{-\delta-2n}/e \end{bmatrix}_k \tag{2.6b}$$

$$\times \quad (-1)^k \frac{[q^{k-1-\delta-2n},\, q^{-\delta-2n};\, q]_k}{(q;\, q)_k} \, q^{k(\varepsilon+\delta-n)-\binom{k}{2}} / (bc)^k \tag{2.6c}$$

$$\times \; _4\phi_3 \begin{bmatrix} q^{-1-\delta-2n+2k}, \, q^{-\delta-2n+k}, & bq^k, & cq^k & \frac{q^{\varepsilon+\delta-n-k}}{bc} \\ q^{-1-\delta-2n+k}, q^{\alpha-2n+k}/b, q^{\gamma-2n+k}/c; \end{bmatrix} \tag{2.6d}$$

Replacing the $_4\phi_3[\cdots]$ by the evaluations listed in Section 1.4, we get the following transforms and the related formulas.

Example 2.19 *For $(\delta;\alpha,\gamma) = (0;0,0)$, the new transforms read as*

$$\Omega_0 \begin{bmatrix} 0,0,0,0; & \varepsilon \\ b,c,d,e; & n \end{bmatrix} \xlongequal{\varepsilon=0,1} \begin{bmatrix} q^{-2n}, & q^{-2n}/bc \\ q^{-2n}/b, & q^{-2n}/c \end{bmatrix}_n$$

$$\times \; _4\phi_3 \begin{bmatrix} q^{-n}, & b, & c, & q^{-2n}/de \\ & q^{-2n}/d, & q^{-2n}/e, & bcq^{n+1} \end{bmatrix}$$

based on the evaluation from Example 1.4

$$_4\phi_3 \begin{bmatrix} q^{-1-2n+2k}, & q^{k-2n}, & bq^k, & cq^k & \frac{q^{\varepsilon-n-k}}{bc} \\ & q^{k-1-2n}, & q^{k-2n}/b, & q^{k-2n}/c \end{bmatrix}$$

$$\xlongequal{\varepsilon=0,1} q^{-\varepsilon k} \frac{1-q^{k-n-1}}{1-q^{k-2n-1}} \begin{bmatrix} q^{-1-2n+2k}, & q^{-2n}/bc \\ q^{k-2n}/b, & q^{k-2n}/c \end{bmatrix}_{n-k}.$$

When $bcde = q^{-1-3n}$, they reduce, by means of the Saalschütz formula (1.1), to the evaluations

$$_5\phi_4 \begin{bmatrix} q^{-2n}, & b, & c, & d, & e \\ & q^{-2n}/b, & q^{-2n}/c, & q^{-2n}/d, & q^{-2n}/e \end{bmatrix}$$

$$\xlongequal{\varepsilon=1,2} \begin{bmatrix} q^{-2n}, & q^{-2n}/bc, & q^{-2n}/bd, & q^{-2n}/cd \\ q^{-2n}/b, & q^{-2n}/c, & q^{-2n}/d, & q^{-2n}/bcd \end{bmatrix}_n.$$

Example 2.20 *For* $(\delta; \alpha, \gamma) = (0; 0, 1)$, *the new transform reads as*

$$\Omega_0 \begin{bmatrix} 0,1,0,0; & 1 \\ b,c,d,e; & n \end{bmatrix} = \begin{bmatrix} q^{-2n}, & q^{1-2n}/bc \\ q^{-2n}/b, & q^{1-2n}/c \end{bmatrix}; q \Big]_n$$

$$\times \quad {}_4\phi_3 \begin{bmatrix} q^{-n}, & b, & c/q, & q^{-2n}/de \\ & q^{-2n}/d, & q^{-2n}/e, & bcq^n \end{bmatrix}; q \Big]$$

based on the evaluation from Example 1.5

$$_4\phi_3 \begin{bmatrix} q^{-1-2n+2k}, & q^{k-2n}, & bq^k, & cq^k \\ & q^{k-1-2n}, & q^{k-2n}/b, & q^{k+1-2n}/c \end{bmatrix}; \frac{q^{1-n-k}}{bc} \Big]$$

$$= \frac{(1-q^{k-n-1})(1-c/q)}{(1-q^{k-2n-1})(1-cq^{k-1})} \begin{bmatrix} q^{-1-2n+2k}, & q^{1-2n}/bc \\ q^{k-2n}/b, & q^{k+1-2n}/c \end{bmatrix}; q \Big]_{n-k}.$$

When $bcde = q^{-3n}$, *it reduces, by means of the Saalschütz formula* (1.1), *to the evaluation*

$$_5\phi_4 \begin{bmatrix} q^{-2n}, & b, & c, & d, & e \\ & q^{-2n}/b, & q^{1-2n}/c, & q^{-2n}/d, & q^{-2n}/e \end{bmatrix}; q \Big]$$

$$= \begin{bmatrix} q^{-2n}, & q^{1-2n}/bc, & q^{-2n}/bd, & q^{1-2n}/cd \\ q^{-2n}/b, & q^{1-2n}/c, & q^{-2n}/d, & q^{1-2n}/bcd \end{bmatrix}; q \Big]_n.$$

Example 2.21 *For* $(\delta; \alpha, \gamma) = (0; 0, -1)$, *the new transform reads as*

$$\Omega_0 \begin{bmatrix} 0,-1,0,0; & 0 \\ b, & c, & d, & e; & n \end{bmatrix} = \begin{bmatrix} q^{-2n}, & q^{-2n}/bc \\ q^{-2n}/b, & q^{-2n}/c \end{bmatrix}; q \Big]_n$$

$$\times \quad {}_4\phi_3 \begin{bmatrix} q^{-n}, & b, & c, & q^{-2n}/de \\ & q^{-2n}/d, & q^{-2n}/e, & bcq^{n+1} \end{bmatrix}; q \Big]$$

based on the evaluation from Example 1.6

$$_4\phi_3 \begin{bmatrix} q^{-1-2n+2k}, & q^{k-2n}, & bq^k, & cq^k \\ & q^{k-1-2n}, & q^{k-2n}/b, & q^{k-1-2n}/c \end{bmatrix}; \frac{q^{-n-k}}{bc} \Big]$$

$$= \frac{(1-q^{k-n-1})(1-q^{-1-2n}/c)}{(1-q^{k-2n-1})(1-q^{-1-n}/c)} \begin{bmatrix} q^{-1-2n+2k}, & q^{-2n}/bc \\ q^{k-2n}/b, & q^{k-1-2n}/c \end{bmatrix}; q \Big]_{n-k}.$$

When $bcde = q^{-1-3n}$, *it reduces, by means of the Saalschütz formula* (1.1), *to*

$$_5\phi_4 \begin{bmatrix} q^{-2n}, & b, & c, & d, & e \\ & q^{-2n}/b, & q^{1-2n}/c, & q^{-2n}/d, & q^{-2n}/e \end{bmatrix}; q \Big]$$

$$= \begin{bmatrix} q^{-2n}, & q^{-2n}/bc, & q^{-2n}/bd, & q^{-2n}/cd \\ q^{-2n}/b, & q^{-2n}/c, & q^{-2n}/d, & q^{-2n}/bcd \end{bmatrix}; q \Big]_n.$$

This is coincident with the evaluation demonstrated in Example 2.19.

In fact, we have a general transform with an extra parameter w

$$_6\phi_5 \begin{bmatrix} q^{-2n}, & qw, & b, & c, & d, & e \\ & w, & q^{-2n}/b, & q^{-2n}/c, & q^{-2n}/d, & q^{-2n}/e \end{bmatrix}; \frac{q^{-3n}}{bcde} \Big] \qquad (2.7a)$$

$$= \begin{bmatrix} q^{-2n}, & q^{-2n}/bc \\ q^{-2n}/b, & q^{-2n}/c \end{bmatrix}; q \Big]_n {}_4\phi_3 \begin{bmatrix} q^{-n}, & b, & c, & q^{-2n}/de \\ & q^{-2n}/d, q^{-2n}/e, & bcq^{n+1} \end{bmatrix}; q \Big] \qquad (2.7b)$$

and its degenerated summation formula (for $bcde = q^{-1-3n}$)

$$
{}_6\phi_5 \left[\begin{array}{cccccc} q^{-2n}, & qw, & b, & c, & d, & e \\ & w, & q^{-2n}/b, & q^{-2n}/c, & q^{-2n}/d, & q^{-2n}/e \end{array} ; q \right] \tag{2.7c}
$$

$$
= \left[\begin{array}{cccc} q^{-2n}, & q^{-2n}/bc, & q^{-2n}/bd, & q^{-2n}/cd \\ q^{-2n}/b, & q^{-2n}/c, & q^{-2n}/d, & q^{-2n}/bcd \end{array} ; q \right]_n \tag{2.7d}
$$

which follow from

$$
{}_6\phi_5 \left[\begin{array}{ccccccc} q^{-2n}, & qw, & b, & c, & d, & e & ; & \dfrac{q^{-3n}}{bcde} \\ & w, & q^{-2n}/b, & q^{-2n}/c, & q^{-2n}/d, & q^{-2n}/e & & \end{array} \right]
$$

$$
= \frac{1}{1-w} \left\{ \Omega_0 \left[\begin{array}{c} 0,0,0,0; \ 0 \\ b,c,d,e; \ n \end{array} \right] - w\,\Omega_0 \left[\begin{array}{c} 0,0,0,0; \ 1 \\ b,c,d,e; \ n \end{array} \right] \right\} = \Omega_0 \left[\begin{array}{c} 0,0,0,0; \ 0 \\ b,c,d,e; \ n \end{array} \right]
$$

in view of Example 2.19.

Example 2.22 *For* $(\delta; \alpha, \gamma) = (1; -1, -1)$, *the new transform reads as*

$$
\Omega_1 \left[\begin{array}{cccc} -1, -1, -1, -1; & -1 \\ b, \quad c, \quad d, \quad e; & n \end{array} \right] = \left[\begin{array}{cc} q^{-1-2n}, & q^{-1-2n}/bc \\ q^{-1-2n}/b, & q^{-1-2n}/c \end{array} ; q \right]_{n+1}
$$

$$
\times \ {}_4\phi_3 \left[\begin{array}{cccc} q^{-1-n}, & b, & c, & q^{-1-2n}/de \\ & q^{-1-2n}/d, & q^{-1-2n}/e, & bcq^{n+1} \end{array} ; q \right]
$$

based on the evaluation from Example 1.2

$$
{}_4\phi_3 \left[\begin{array}{cccc} q^{-2-2n+2k}, & q^{k-1-2n}, & bq^k, & cq^k \\ q^{k-2-2n}, & q^{k-1-2n}/b, & q^{k-1-2n}/c \end{array} ; \frac{q^{-n-k}}{bc} \right]
$$

$$
= \frac{1-q^{-1-n}}{1-q^{k-2-2n}} \left[\begin{array}{cc} q^{-2-2n+2k}, q^{-1-2n}/bc \\ q^{k-1-2n}/b, q^{k-1-2n}/c \end{array} ; q \right]_{1+n-k}.
$$

When $bcde = q^{-2-3n}$, *it reduces, by means of the Saalschütz formula* (1.1), *to the evaluation*

$$
{}_5\phi_4 \left[\begin{array}{ccccc} q^{-1-2n}, & b, & c, & d, & e \\ & q^{-1-2n}/b, & q^{-1-2n}/c, & q^{-1-2n}/d, & q^{-1-2n}/e \end{array} ; q \right]
$$

$$
= \left[\begin{array}{cccc} q^{-1-2n}, & q^{-1-2n}/bc, & q^{-1-2n}/bd, & q^{-1-2n}/cd \\ q^{-1-2n}/b, & q^{-1-2n}/c, & q^{-1-2n}/d, & q^{-1-2n}/bcd \end{array} ; q \right]_{n+1}.
$$

Example 2.23 *For* $(\delta; \alpha, \gamma) = (1; -2, -2)$, *the new transform reads as*

$$
\Omega_1 \left[\begin{array}{cccc} -2, -2, -1, -1; & -2 \\ b, \quad c, \quad d, \quad e; & n \end{array} \right] = \left[\begin{array}{cc} q^{-2-2n}, & q^{-2-2n}/bc \\ q^{-2-2n}/b, & q^{-2-2n}/c \end{array} ; q \right]_{n+1}
$$

$$
\times \ {}_5\phi_4 \left[\begin{array}{ccccc} q^{-1-n}, q^{-1-2n}, & b, & c, & q^{-1-2n}/de \\ & q^{-2-2n}, & q^{-1-2n}/d, q^{-1-2n}/e, & bcq^{n+2} \end{array} ; q \right]
$$

based on the evaluation from Example 1.1

$$
{}_4\phi_3 \left[\begin{array}{cccc} q^{-2-2n+2k}, & q^{k-1-2n}, & bq^k, & cq^k \\ q^{k-2-2n}, & q^{k-2-2n}/b, & q^{k-2-2n}/c \end{array} ; \frac{q^{-1-n-k}}{bc} \right]
$$

$$
= \left[\begin{array}{cc} q^{-2-2n+2k}, q^{-2-2n}/bc \\ q^{k-2-2n}/b, q^{k-2-2n}/c \end{array} ; q \right]_{1+n-k}.
$$

When $bcde = q^{-3-3n}$, *it reduces, by means of the 2-balanced separation (2.4) via substituting* $w \to q^{-2-2n}$, *to the evaluation*

$$
{}_5\phi_4 \begin{bmatrix} q^{-1-2n}, & b, & c, & d, & e \\ & q^{-2-2n}/b, & q^{-2-2n}/c, & q^{-1-2n}/d, & q^{-1-2n}/e \end{bmatrix} ; q \end{bmatrix}
$$

$$
= \begin{bmatrix} q^{-1-2n}, & q^{-2-2n}/bc, & q^{-2-2n}/bd, & q^{-2-2n}/cd \\ q^{-2-2n}/b, & q^{-2-2n}/c, & q^{-1-2n}/d, & q^{-2-2n}/bcd \end{bmatrix} ; q \end{bmatrix}_{n+1}
$$

$$
\times \left\{ 1 + q^{1+n} \begin{bmatrix} d, & bcdq^{2n+2} \\ bdq^{2n+2}, & cdq^{2n+2} \end{bmatrix} \right\}.
$$

Remark For $(\delta; \alpha, \gamma) = (1; 0, 0)$, the corresponding transform reads as, based on the evaluation from Example 1.3,

$$
\Omega_1 \begin{bmatrix} 0, 0, -1, -1; & 0 \\ b, c, d, & e; & n \end{bmatrix} = \begin{bmatrix} q^{-1-n}, & q^{-1-2n}/de \\ q^{-1-2n}/d, & q^{-1-2n}/e \end{bmatrix} \times \begin{bmatrix} q^{-1-2n}, & q^{-2n}/bc \\ q^{-2n}/b, & q^{-2n}/c \end{bmatrix} ; q \end{bmatrix}_n
$$

$$
\times {}_4\phi_3 \begin{bmatrix} q^{-n}, & b, & c, & q^{-2n}/de \\ & q^{-2n}/d, & q^{-2n}/e, & bcq^{n+1} \end{bmatrix} ; q \end{bmatrix}
$$

which is equivalent, in view of the Hall formula (2.2c), to the transformation established in Example 2.14.

2.3   Case $(1 + \beta = \rho = 1 - \delta)$

Taking

$$
\begin{aligned}
1 + \beta = \rho &= 1 - \delta \\
\lambda = \mu = \nu &= 1 - \delta
\end{aligned}
$$

in the general relation (2.3), the following transform may be established after some trivial modification.

Proposition III

$$
\Omega_\delta \begin{bmatrix} \alpha, & \gamma, & -\delta, & 1 - \delta; & \varepsilon \\ b, & c, & d, & e; & n \end{bmatrix} \tag{2.8a}
$$

$$
= \sum_k \begin{bmatrix} b, & c, & q^{1-\delta-2n}/de \\ q^{\alpha-2n}/b, & q^{\gamma-2n}/c, & q^{-\delta-2n}/d, \ q^{1-\delta-2n}/e \end{bmatrix}_k \tag{2.8b}
$$

$$
\times \quad (-1)^k \ \frac{(q^{-\delta-2n}; q)_{2k}}{(q; q)_k} \ q^{k(\varepsilon+\delta-1-n)-\binom{k}{2}} / (bc)^k \tag{2.8c}
$$

$$
\times \ {}_4\phi_3 \begin{bmatrix} q^{-\delta-2n+2k}, q^{1-\delta-2n+k}/d, & bq^k, & cq^k & q^{\varepsilon+\delta-1-n-k} \\ & q^{-\delta-2n+k}/d, \ q^{\alpha-2n+k}/b, \ q^{\gamma-2n+k}/c & ; & bc \end{bmatrix} \tag{2.8d}
$$

Replacing the $_4\phi_3[\cdots]$ by the evaluations listed in Section 1.4, we have the following transforms and the related formulas.

Example 2.24 *For* $(\delta; \alpha, \gamma) = (0; 2, 2)$, *the new transform reads as*

$$
\Omega_0 \begin{bmatrix} 2, 2, 0, 1; & 3 \\ b, c, d, e; & n \end{bmatrix} = q^n \begin{bmatrix} bq^{n-1}, cq^{n-1}, d \\ b/q, c/q, dq^{2n} \end{bmatrix} \times \begin{bmatrix} q^{-2n}, & q^{2-2n}/bc \\ q^{2-2n}/b, & q^{2-2n}/c \end{bmatrix}_n
$$

$$
\times \ {}_5\phi_4 \begin{bmatrix} q^{-n}, & qd, & b/q, & c/q, & q^{1-2n}/de \\ & d, & q^{1-2n}/d, & q^{1-2n}/e, & bcq^{n-1} \end{bmatrix} ; q
$$

based on the evaluation from Example 1.3

$$
{}_4\phi_3 \begin{bmatrix} q^{-2n+2k}, & q^{k+1-2n}/d, & bq^k, & cq^k & q^{2-n-k} \\ & q^{k-2n}/d, & q^{k+2-2n}/b, & q^{k+2-2n}/c & ; & bc \end{bmatrix}
$$

$$
= \frac{(1 - bq^{n-1})(1 - cq^{n-1})(1 - q^{-k}/d)}{(1 - bq^{k-1})(1 - cq^{k-1})(1 - q^{k-2n}/d)} \begin{bmatrix} q^{2k-2n}, & q^{2-2n}/bc \\ q^{k+2-2n}/b, q^{k+2-2n}/c \end{bmatrix}_{n-k}.
$$

When $bcde = q^{2-3n}$, *it reduces, by means of the 2-balanced separation (2.4) via substituting* $w \to d$, *to the evaluation*

$$
{}_5\phi_4 \begin{bmatrix} q^{-2n}, & b, & c, & d, & e \\ & q^{2-2n}/b, & q^{2-2n}/c, & q^{-2n}/d, & q^{1-2n}/e \end{bmatrix} ; q
$$

$$
= q^{-n} \begin{bmatrix} bq^{n-1}, cq^{n-1} \\ b/q, c/q \end{bmatrix} \times \left\{ 1 - d \begin{bmatrix} q^n, & bcq^{n-2} \\ bdq^{n-1}, cdq^{n-1} \end{bmatrix} \right\}
$$

$$
\times \begin{bmatrix} q^{-2n}, & q^{2-2n}/bc, & q^{2-2n}/bd, & q^{2-2n}/cd \\ q^{2-2n}/b, & q^{2-2n}/c, & q^{-2n}/d, & q^{2-2n}/bcd \end{bmatrix}_n.
$$

Example 2.25 *For* $(\delta; \alpha, \gamma) = (1; -1, -1)$, *let*

$$u = 1 - bq^{n+1} - cq^{n+1} + bcdq^{3n+2},$$

$$v = 1 - bdq^{2n+1} - cdq^{2n+1} + bcdq^{3n+2}.$$

The new transforms read as

$$\Omega_1 \begin{bmatrix} -1, -1, -1, 0; \ \varepsilon \\ b, \quad c, \quad d, \quad e; \ n \end{bmatrix} \overset{\varepsilon = 0, -1}{=\!=\!=\!=} q^{-n}/d \begin{bmatrix} dq^n \\ bcq^{n+1}, q^{-1-2n}/d \end{bmatrix} \begin{bmatrix} q^{-1-2n}, \ q^{-1-2n}/bc \\ q^{-1-2n}/b, \ q^{-1-2n}/c \end{bmatrix}_{n+1}$$

$$\times \begin{cases} -\{1 + \frac{q^{-1-n}u}{1-dq^n}\} \ _5\phi_4 \begin{bmatrix} q^{-n}, \ \frac{-q^{-n}u}{1-dq^n}, \quad b, \quad c, \quad q^{-2n}/de \\ \frac{-q^{-1-n}u}{1-dq^n}, q^{-2n}/d, q^{-2n}/e, bcq^{n+2} \end{bmatrix} & (\varepsilon = 0) \\[3mm] bc\{1 - \frac{q^{-1-n}v}{bc(1-dq^n)}\} \ _5\phi_4 \begin{bmatrix} q^{-n}, q^{2+n}bc\frac{1-dq^n}{v}, \quad b, \quad c, \quad q^{-2n}/de \\ q^{1+n}bc\frac{1-dq^n}{v}, q^{-2n}/d, q^{-2n}/e, bcq^{n+2} \end{bmatrix} & (\varepsilon = -1) \end{cases}$$

based on the computation from Section 1.3

$$_4\phi_3 \begin{bmatrix} q^{-1-2n+2k}, \quad q^{k-2n}/d \quad bq^k, \quad cq^k \\ q^{k-1-2n}/d, \ q^{k-1-2n}/b, \ q^{k-1-2n}/c \end{bmatrix} ; \ \frac{q^{\varepsilon-n-k}}{bc} \end{bmatrix}$$

$$= \frac{1}{1-q^{k-1-2n}/d} \left\{ \omega_1 \begin{bmatrix} -1, -1; \ \varepsilon \\ bq^k, cq^k; n-k \end{bmatrix} - q^{k-1-2n}/d \, \omega_1 \begin{bmatrix} -1, -1; 1+\varepsilon \\ bq^k, cq^k; n-k \end{bmatrix} \right\}$$

$$= \begin{bmatrix} dq^n \\ bcq^{k+1+n}, q^{k-1-2n}/d \end{bmatrix} \begin{bmatrix} q^{-1-2n+2k}, \quad q^{-1-2n}/bc \\ q^{-1-2n+k}/b, \ q^{-1-2n+k}/c \end{bmatrix}_{1+n-k}$$

$$\times \begin{cases} -q^{-n}/d \ \{1 + q^{k-1-n}u/(1-dq^n)\} & (\varepsilon = 0) \\ q^{2k-n}bc/d \ \{1 - b^{-1}c^{-1}q^{-1-n-k}v/(1-dq^n)\} & (\varepsilon = -1). \end{cases}$$

When $bcde = q^{-2-3n}$, *they reduce, by means of the 2-balanced separation* (2.4) *via substituting* $w \to -q^{-1-n}u/(1-dq^n)$, $bcq^{1+n}(1-dq^n)/v$ *in accordance with* $\varepsilon = 0, -1$, *respectively, to evaluations*

$$_5\phi_4 \begin{bmatrix} q^{-2n}, \quad b, \quad c, \quad d, \quad e \\ q^{-1-2n}/b, \ q^{-1-2n}/c, \ q^{-1-2n}/d \ q^{-2n}/e \end{bmatrix} ; \ q^\varepsilon \end{bmatrix}$$

$$\overset{\varepsilon=1,2}{=\!=\!=} \begin{bmatrix} q^{-1-2n}, \ q^{-1-2n}/bc \\ q^{-1-2n}/b, \ q^{-1-2n}/c \end{bmatrix}_{n+1} \begin{bmatrix} q^{-1-2n}/bd, \ q^{-1-2n}/cd \\ q^{-1-2n}/d, \ q^{-1-2n}/bcd \end{bmatrix}_n$$

$$\times \quad q^{(\varepsilon-1)(n+1)} \frac{1-d}{1-dq^{n+1}} \times \left\{ 1 + q^{(1-\varepsilon)(n+1)} \, d^{2-\varepsilon} \begin{bmatrix} bq^{n+1}, cq^{n+1} \\ d, bcq^{n+1} \end{bmatrix} \right\}.$$

Example 2.26 *For* $(\delta; \alpha, \gamma) = (1; -2, -1)$, *let*

$$u = -bq - q^{-n} + bcq^{n+2} + q^{-2n}/d,$$

$$v = bcq + q^{-1-3n}/d - q^{-2n}c/d - q^{1-n}bc/d.$$

The new transform reads as

$$
\Omega_1 \begin{bmatrix} -2, -1, -1, 0; & -1 \\ b, & c, & d, & e; & n \end{bmatrix} = \begin{bmatrix} q^{-1-n} \\ bq^{n+2}, cq^{n+1}, q^{-1-2n}/d \end{bmatrix} \begin{bmatrix} q^{-1-2n}, q^{-1-2n}/bc \\ q^{-2-2n}/b, q^{-1-2n}/c \end{bmatrix}; q \Big]_n
$$

$$
\times \quad (u+v) \ q^{n+1} \ {}_5\phi_4 \begin{bmatrix} q^{-n}, & -qv/u, & b, & c, & q^{-2n}/de \\ & -v/u, & q^{-2n}/d, & q^{-2n}/e, & bcq^{n+2} \end{bmatrix}; q \Big]
$$

based on the computation from Section 1.3

$$
{}_4\phi_3 \begin{bmatrix} q^{-1-2n+2k}, & q^{k-2n}/d & bq^k, & cq^k & q^{-1-n-k} \\ & q^{k-1-2n}/d, & q^{k-2-2n}/b, & q^{k-1-2n}/c & ; \quad \dfrac{q^{-1-n-k}}{bc} \end{bmatrix}
$$

$$
= \frac{1}{1-q^{k-1-2n}/d} \left\{ \omega_1 \begin{bmatrix} -2, -1; & -1 \\ bq^k, cq^k; n-k \end{bmatrix} - q^{k-1-2n}/d \ \omega_1 \begin{bmatrix} -2, -1; & 0 \\ bq^k, cq^k; n-k \end{bmatrix} \right\}
$$

$$
= q^{n+1} (u+vq^k) \begin{bmatrix} q^{k-n-1} \\ bq^{n+2}, cq^{n+1}, q^{k-1-2n}/d \end{bmatrix} \begin{bmatrix} q^{-1-2n+2k}, & q^{-1-2n}/bc \\ q^{-2-2n+k}/b, & q^{-1-2n+k}/c \end{bmatrix}; q \Big]_{n-k}.
$$

When $bcde = q^{-2-3n}$, it reduces, by means of the 2-balanced separation (2.4) via substituting $w \to -v/u$, to the evaluation

$$
{}_5\phi_4 \begin{bmatrix} q^{-2n}, & b, & c, & d, & e \\ & q^{-2-2n}/b, & q^{-1-2n}/c, & q^{-1-2n}/d & q^{-2n}/e \end{bmatrix}; q \Big]
$$

$$
= \ q^{-n} \begin{bmatrix} q^{-1-2n}, & q^{-1-2n}/bc, & q^{-1-2n}/bd, & q^{-1-2n}/cd \\ q^{-2-2n}/b, & q^{-1-2n}/c, & q^{-1-2n}/d, & q^{-1-2n}/bcd \end{bmatrix}; q \Big]_n
$$

$$
\times \left\{ 1 + q^{n+1} \begin{bmatrix} c, bdq^{n+1} \\ bcq^{2n+2}, dq^{n+1} \end{bmatrix} \right\} \begin{bmatrix} q^{n+1}, & bcq^{2n+2} \\ bq^{n+2}, & cq^{n+1} \end{bmatrix}.
$$

Remark

a. For $(\delta; \alpha, \gamma) = (0; 0, 0)$, the corresponding transform reads as, based on the evaluation from Example 1.1,

$$
\Omega_0 \begin{bmatrix} 0, 0, 0, 1; & 1 \\ b, c, d, e; & n \end{bmatrix} = \begin{bmatrix} q^{-2n}, & q^{-2n}/bc \\ q^{-2n}/b, & q^{-2n}/c \end{bmatrix}; q \Big]_n
$$

$$
\times \ {}_4\phi_3 \begin{bmatrix} q^{-n}, & b, & c, & q^{1-2n}/de \\ & q^{-2n}/d, & q^{1-2n}/e, & bcq^{n+1} \end{bmatrix}; q \Big]
$$

which is equivalent, in view of the Sears formula (2.2b), to the transformation established in Example 2.20.

b. For $(\delta; \alpha, \gamma) = (0; 1, 1)$, the corresponding transform reads as, based on the evaluation from Example 1.2,

$$
\Omega_0 \begin{bmatrix} 1, 1, 0, 1; & 2 \\ b, c, d, e; & n \end{bmatrix} = \begin{bmatrix} q^{-2n}, & q^{1-2n}/bc, & q^{1-2n}/d \\ q^{1-2n}/b, & q^{1-2n}/c, & q^{-2n}/d \end{bmatrix}; q \Big]_n
$$

$$
\times \ {}_4\phi_3 \begin{bmatrix} q^{-n}, & b, & c, & q^{1-2n}/de \\ & q^{1-2n}/d, & q^{1-2n}/e, & bcq^n \end{bmatrix}; q \Big]
$$

which is equivalent, in view of the Sears formula (2.2b), to the transformation established in Example 2.4.

c. For $(\delta; \alpha, \gamma) = (1; 0, 0)$, the corresponding transforms read as, based on the evaluation from Example 1.4,

$$\Omega_1 \begin{bmatrix} 0,0,-1,0; & \varepsilon \\ b,c,d,e; & n \end{bmatrix} \overset{\varepsilon=0,1}{=\!=\!=} d^{-\varepsilon} \begin{bmatrix} q^{-1-n} \\ q^{-1-2n}/d \end{bmatrix} \times \begin{bmatrix} q^{-1-2n}, q^{-2n}/bc \\ q^{-2n}/b, \ q^{-2n}/c \end{bmatrix} \ ; \ q \Big]_n$$
$$\times \ _4\phi_3 \begin{bmatrix} q^{-n}, & b, & c, & q^{-2n}/de \\ & q^{-2n}/d, & q^{-2n}/e, & bcq^{n+1} \end{bmatrix} \ ; \ q \Big]$$

which is equivalent, in view of the Sears formula (2.2b), to the transformation established in Example 2.12.

d. For $(\delta; \alpha, \gamma) = (1; 0, 1)$, the corresponding transform reads as, based on the evaluation from Example 1.5,

$$\Omega_1 \begin{bmatrix} 0,1,-1,0; & 1 \\ b,c,d,e; & n \end{bmatrix} = q^n \begin{bmatrix} q^{1+n}, \ qd/c \\ dq^{1+2n}, \ q/c \end{bmatrix} \times \begin{bmatrix} q^{-1-2n}, & q^{1-2n}/bc \\ q^{-2n}/b, & q^{1-2n}/c \end{bmatrix} \ ; \ q \Big]_n$$
$$\times \ _4\phi_3 \begin{bmatrix} q^{-n}, & b, & c/q, & q^{-2n}/de \\ & q^{-2n}/d, & q^{-2n}/e, & bcq^n \end{bmatrix} \ ; \ q \Big]$$

which is equivalent, in view of the Sears formula (2.2b), to the transformation established in Example 2.10.

e. For $(\delta; \alpha, \gamma) = (1; 0, -1)$, the corresponding transform reads as, based on the evaluation from Example 1.6,

$$\Omega_1 \begin{bmatrix} 0,-1,-1,0; & 0 \\ b, \ c, \ d, \ e; & n \end{bmatrix} = \begin{bmatrix} q^{1+n}, \ cdq^{1+2n} \\ cq^{1+n}, \ dq^{1+2n} \end{bmatrix} \times \begin{bmatrix} q^{-1-2n}, \ q^{-2n}/bc \\ q^{-2n}/b, \ q^{1-2n}/c \end{bmatrix} \ ; \ q \Big]_n$$
$$\times \ _4\phi_3 \begin{bmatrix} q^{-n}, & b, & c, & q^{-2n}/de \\ & q^{-2n}/d, & q^{-2n}/e, & bcq^{n+1} \end{bmatrix} \ ; \ q \Big]$$

which is equivalent, in view of the Sears formula (2.2b), to the transformation established in Example 2.14.

2.4   Case $(\beta - 1 = \rho = 1 - \delta)$

Taking

$$\beta - 1 = \rho = 1 - \delta$$
$$\lambda = \mu - 1 = \nu = 1 - \delta$$

in the general relation (2.3), the following transform may be established after some trivial modification.

Proposition IV

$$\Omega_\delta \begin{bmatrix} \alpha, & \gamma, & 2 - \delta, & 1 - \delta; & \varepsilon \\ b, & c, & d, & e; & n \end{bmatrix} \tag{2.9a}$$

$$= \sum_k \begin{bmatrix} b, & c, & d, & q^{2-\delta-2n}/de \\ q^{\alpha-2n}/b, & q^{\gamma-2n}/c, & d/q, & q^{2-\delta-2n}/d, & q^{1-\delta-2n}/e \end{bmatrix} ; q \end{bmatrix}_k \tag{2.9b}$$

$$\times \quad (-1)^k \quad \frac{(q^{-\delta-2n}; q)_{2k}}{(q; q)_k} \quad q^{k(\varepsilon+\delta-2-n)-\binom{k}{2}} / (bc)^k \tag{2.9c}$$

$$\times \, {}_4\phi_3 \begin{bmatrix} q^{-\delta-2n+2k}, & q^k d, & bq^k, & cq^k; & q^{\varepsilon+\delta-2-n-k} \\ & q^{k-1}d, & q^{\alpha-2n+k}/b, & q^{\gamma-2n+k}/c; & bc \end{bmatrix} \tag{2.9d}$$

Replacing the ${}_4\phi_3[\cdots]$ by the evaluations listed in Section 1.4, we find the following transforms and the related formulas.

Example 2.27 *For $(\delta; \alpha, \gamma) = (0; 0, 0)$, the new transform reads as*

$$\Omega_0 \begin{bmatrix} 0, 0, 2, 1; & 2 \\ b, c, d, e; & n \end{bmatrix} = \begin{bmatrix} q^{-2n}, & q^{-2n}/bc \\ q^{-2n}/b, & q^{-2n}/c \end{bmatrix} ; q \end{bmatrix}_n$$

$$\times \, {}_5\phi_4 \begin{bmatrix} q^{-n}, & d, & b, & c, & q^{2-2n}/de \\ & d/q, & q^{2-2n}/d, & q^{1-2n}/e, & bcq^{n+1} \end{bmatrix} ; q \end{bmatrix}$$

based on the evaluation from Example 1.1

$${}_4\phi_3 \begin{bmatrix} q^{-2n+2k}, & q^k d, & bq^k, & cq^k; & q^{-n-k} \\ & q^{k-1}d, & q^{k-2n}/b, & q^{k-2n}/c; & bc \end{bmatrix}$$

$$= \begin{bmatrix} q^{2k-2n}, & q^{-2n}/bc \\ q^{k-2n}/b, & q^{k-2n}/c \end{bmatrix} ; q \end{bmatrix}_{n-k}.$$

When $bcde = q^{1-3n}$, it reduces, by means of the 2-balanced separation (2.4) via substituting $w \to d/q$, to the evaluation

$${}_5\phi_4 \begin{bmatrix} q^{-2n}, & b, & c, & d, & e \\ & q^{-2n}/b, & q^{-2n}/c, & q^{2-2n}/d, & q^{1-2n}/e \end{bmatrix} ; q \end{bmatrix}$$

$$= q^{-n} \begin{bmatrix} q^{-2n}, & q^{-2n}/bc, & q^{2-2n}/bd, & q^{2-2n}/cd \\ q^{-2n}/b, & q^{-2n}/c, & q^{2-2n}/d, & q^{1-2n}/bcd \end{bmatrix} ; q \end{bmatrix}_n$$

$$\times \begin{bmatrix} dq^{n-1} \\ d/q \end{bmatrix} \times \left\{ 1 - d/q \begin{bmatrix} q^n, & bcq^n \\ bdq^{n-1}, cdq^{n-1} \end{bmatrix} \right\}.$$

Example 2.28 *For $(\delta; \alpha, \gamma) = (0; 2, 2)$, the new transform reads as*

$$\Omega_0 \begin{bmatrix} 2,2,2,1; & 4 \\ b,c,d,e; & n \end{bmatrix} = q^{-n} \begin{bmatrix} bq^{n-1}, cq^{n-1}, dq^{2n-1} \\ b/q, \quad c/q, \quad d/q \end{bmatrix} \times \begin{bmatrix} q^{-2n}, q^{2-2n}/bc \\ q^{2-2n}/b, q^{2-2n}/c \end{bmatrix}; q \end{bmatrix}_n$$

$$\times {}_4\phi_3 \begin{bmatrix} q^{-n}, & b/q, & c/q, & q^{2-2n}/de \\ & q^{1-2n}/d, & q^{1-2n}/e, & bcq^{n-1} \end{bmatrix}; q \end{bmatrix}$$

based on the evaluation from Example 1.3

$$ {}_4\phi_3 \begin{bmatrix} q^{-2n+2k}, & q^k d, & bq^k, & cq^k & ; & \dfrac{q^{2-n-k}}{bc} \\ & q^{k-1}d, & q^{k+2-2n}/b, & q^{k+2-2n}/c & & \end{bmatrix}$$

$$ = \dfrac{(1-bq^{n-1})(1-cq^{n-1})(1-q^{2n-k-1}d)}{(1-bq^{k-1})(1-cq^{k-1})(1-q^{k-1}d)} \begin{bmatrix} q^{2k-2n}, & q^{2-2n}/bc \\ q^{k+2-2n}/b, q^{k+2-2n}/c \end{bmatrix}; q \end{bmatrix}_{n-k}. $$

When $bcde = q^{3-3n}$, it reduces, by means of the Saalschütz formula (1.1), to the evaluation

$$ {}_5\phi_4 \begin{bmatrix} q^{-2n}, & b, & c, & d, & e & ; q \\ & q^{2-2n}/b, & q^{2-2n}/c, & q^{2-2n}/d, & q^{1-2n}/e & \end{bmatrix}$$

$$ = \begin{bmatrix} b, & c, & d, & q^{-2n}, & q^{2-2n}/bc, q^{2-2n}/bd, q^{2-2n}/cd \\ b/q, c/q, d/q, & q^{2-2n}/b, & q^{2-2n}/c, & q^{2-2n}/d, & q^{3-2n}/bcd \end{bmatrix}; q \end{bmatrix}_n. $$

Example 2.29 *For $(\delta; \alpha, \gamma) = (1; 1, 1)$, let*

$$ u = 1 - bq^n - cq^n + bcdq^{3n-1}, $$

$$ v = 1 - bdq^{2n-1} - cdq^{2n-1} + bcdq^{3n-1}. $$

The new transforms read as

$$\Omega_1 \begin{bmatrix} 1,1,1,0; & \varepsilon \\ b,c,d,e; & n \end{bmatrix} \stackrel{\varepsilon = 2,3}{=\!=\!=} (-bcq^{2n-1})^{\varepsilon-2} \begin{bmatrix} q^{-1-n}, dq^{n-1} \\ b/q, c/q, d/q \end{bmatrix} \begin{bmatrix} q^{-1-2n}, q^{1-2n}/bc \\ q^{1-2n}/b, q^{1-2n}/c \end{bmatrix}; q \end{bmatrix}_n$$

$$\times \left\{ \begin{aligned} & \left\{ 1 + \dfrac{q^{-1-n}u}{1-dq^{n-1}} \right\} {}_5\phi_4 \begin{bmatrix} q^{-n}, & \dfrac{-q^{-n}u}{1-dq^{n-1}}, & b/q, & c/q, & q^{1-2n}/de \\ & \dfrac{-q^{-1-n}u}{1-dq^{n-1}}, & q^{1-2n}/d, q^{-2n}/e, & bcq^n \end{bmatrix}; q \end{bmatrix} \quad (\varepsilon=2) \\[2em] & \left\{ 1 - \dfrac{q^{1-n}v}{bc(1-dq^{n-1})} \right\} {}_5\phi_4 \begin{bmatrix} q^{-n}, q^n bc\dfrac{1-dq^{n-1}}{v}, & b/q, & c/q, & q^{1-2n}/de \\ q^n bc\dfrac{1-dq^{n-1}}{qv}, & q^{1-2n}/d, q^{-2n}/e, & bcq^n \end{bmatrix}; q \end{bmatrix} \quad (\varepsilon=3) \end{aligned} \right.$$

based on the computation from Section 1.3

$$ {}_4\phi_3 \begin{bmatrix} q^{-1-2n+2k}, & q^k d, & bq^k, & cq^k & ; & \dfrac{q^{\varepsilon-1-n-k}}{bc} \\ & q^{k-1}d, q^{k+1-2n}/b, q^{k+1-2n}/c & & \end{bmatrix}$$

$$ = \dfrac{(1-bdq^{2n-1})(1-q^{k-2n}/c)}{(1-b/c)(1-dq^{k-1})} \omega_1 \begin{bmatrix} 1, & 0; & \varepsilon-1 \\ bq^k, & cq^k; & n-k \end{bmatrix}$$

$$ + \dfrac{(1-cdq^{2n-1})(1-q^{k-2n}/b)}{(1-c/b)(1-dq^{k-1})} \omega_1 \begin{bmatrix} 0, & 1; & \varepsilon-1 \\ bq^k, & cq^k; & n-k \end{bmatrix}$$

$$= \begin{bmatrix} q^{k-1-n} \\ bq^{k-1}, cq^{k-1}, dq^{k-1} \end{bmatrix} \begin{bmatrix} q^{-1-2n+2k}, & q^{1-2n}/bc \\ q^{1-2n+k}/b, & q^{1-2n+k}/c \end{bmatrix}; q \end{bmatrix}_{n-k}$$

$$\times \begin{cases} q^{k-n-1}u\{1 + q^{1+n-k}u^{-1}(1 - dq^{n-1})\} & (\varepsilon = 2) \\ q^{n-k}v\{1 - q^{k-1+n}bcv^{-1}(1 - dq^{n-1})\} & (\varepsilon = 3). \end{cases}$$

When $bcde = q^{1-3n}$, they reduce, by means of the 2-balanced separation (2.4) via substituting $w \to -q^{-1-n}u/(1 - dq^{n-1}), bcq^{n-1}(1 - dq^{n-1})/v$ in accordance with $\varepsilon = 2, 3$, respectively, to evaluations

$$_5\phi_4 \begin{bmatrix} q^{-2n}, & b, & c, & d, & e \\ & q^{1-2n}/b, & q^{1-2n}/c, & q^{1-2n}/d, & q^{-2n}/e \end{bmatrix}; q^\varepsilon \end{bmatrix}$$

$$\overset{\varepsilon=1,2}{=\!=\!=} q^\varepsilon \begin{bmatrix} q^{-1-2n}, & q^{1-2n}/bc, & q^{1-2n}/bd, & q^{1-2n}/cd \\ q^{1-2n}/b, & q^{1-2n}/c, & q^{1-2n}/d, & q^{2-2n}/bcd \end{bmatrix}; q \end{bmatrix}_n$$

$$\times \begin{bmatrix} q^{-1-n}, bcq^{n-1} \\ b/q, & c/q \end{bmatrix} \times \left\{ 1 + q^{-1+(\varepsilon-2)n} d^{\varepsilon-1} \begin{bmatrix} bq^n, cq^n \\ d/q, bcq^{n-1} \end{bmatrix} \right\}.$$

Example 2.30 *For $(\delta; \alpha, \gamma) = (1; 2, 1)$, let*

$$u = 1 - dq^{n-1} - bdq^{2n-2} + bcdq^{3n-2},$$

$$v = q^{-1-n} - c/q - bcq^{n-2} + bcdq^{2n-3}.$$

The new transform reads as

$$\Omega_1 \begin{bmatrix} 2,1,1,0; 3 \\ b,c,d,e; n \end{bmatrix} = \begin{bmatrix} q^{-1-n}, & bq^{n-1} \\ b/q, b/q^2, c/q, d/q \end{bmatrix} \begin{bmatrix} q^{-1-2n}, & q^{2-2n}/bc \\ q^{2-2n}/b, & q^{1-2n}/c \end{bmatrix}; q \end{bmatrix}_n$$

$$\times (u+v) \; _5\phi_4 \begin{bmatrix} q^{-n}, & -qv/u, & b/q^2, & c/q, & q^{1-2n}/de \\ & -v/u, & q^{1-2n}/d, & q^{-2n}/e, & bcq^{n-1} \end{bmatrix}; q \end{bmatrix}$$

based on the computation from Section 1.3

$$_4\phi_3 \begin{bmatrix} q^{-1-2n+2k}, & q^k d & bq^k, & cq^k \\ & q^{k-1}d, & q^{k+2-2n}/b, & q^{k+1-2n}/c \end{bmatrix}; \frac{q^{2-n-k}}{bc} \end{bmatrix}$$

$$= \frac{1}{1 - dq^{k-1}} \left\{ \omega_1 \begin{bmatrix} 2, & 1; & 2 \\ bq^k, cq^k; n-k \end{bmatrix} - dq^{k-1} \, \omega_1 \begin{bmatrix} 2, & 1; & 3 \\ bq^k, cq^k; n-k \end{bmatrix} \right\}$$

$$= (u+vq^k) \begin{bmatrix} q^{k-n-1}, bq^{n-1} \\ bq^{k-1}, bq^{k-2}, cq^{k-1}, dq^{k-1} \end{bmatrix} \begin{bmatrix} q^{-1-2n+2k}, & q^{2-2n}/bc \\ q^{2-2n+k}/b, & q^{1-2n+k}/c \end{bmatrix}; q \end{bmatrix}_{n-k}.$$

When $bcde = q^{2-3n}$, it reduces, by means of the 2-balanced separation (2.4) via substituting $w \to -v/u$, to the evaluation

$$_5\phi_4 \begin{bmatrix} q^{-2n}, & b, & c, & d, & e \\ & q^{2-2n}/b, & q^{1-2n}/c, & q^{1-2n}/d & q^{-2n}/e \end{bmatrix}; q \end{bmatrix}$$

$$= q^n \begin{bmatrix} q^{-1-2n}, & q^{2-2n}/bc, & q^{2-2n}/bd, & q^{1-2n}/cd \\ q^{2-2n}/b, & q^{1-2n}/c, & q^{1-2n}/d, & q^{3-2n}/bcd \end{bmatrix}; q \end{bmatrix}_n$$

$$\times \left\{ 1 + q^{-1-n} \begin{bmatrix} cq^n, bdq^{2n-1} \\ d/q, bcq^{n-2} \end{bmatrix} \right\} \begin{bmatrix} q^{-1-n}, bq^{n-1}, bcq^{n-2} \\ b/q, & b/q^2, & c/q \end{bmatrix}.$$

Remark

a. For $(\delta; \alpha, \gamma) = (0; 1, 1)$, the corresponding transform reads as, based on the evaluation from Example 1.2,

$$\Omega_0 \begin{bmatrix} 1,1,2,1; & 3 \\ b,c,d,e; & n \end{bmatrix} = \begin{bmatrix} dq^{n-1} \\ d/q \end{bmatrix} \times \begin{bmatrix} q^{-2n}, & q^{1-2n}/bc \\ q^{1-2n}/b, & q^{1-2n}/c \end{bmatrix} ; q \Big]_n$$
$$\times \, {}_4\phi_3 \begin{bmatrix} q^{-n}, & b, & c, & q^{2-2n}/de \\ & q^{2-2n}/d, & q^{1-2n}/e, & bcq^n \end{bmatrix} ; q \Big]$$

which is equivalent, in view of the Sears formula (2.2b), to the transformation established in Example 2.3.

b. For $(\delta; \alpha, \gamma) = (1; 0, 0)$, the corresponding transforms read as, based on the evaluation from Example 1.4,

$$\Omega_1 \begin{bmatrix} 0,0,1,0; & \varepsilon \\ b,c,d,e; & n \end{bmatrix} \overset{\varepsilon=1,2}{=\!=\!=\!=} (dq^{2n})^{\varepsilon-1} \begin{bmatrix} q^{-1-n} \\ d/q \end{bmatrix} \times \begin{bmatrix} q^{-1-2n}, q^{-2n}/bc \\ q^{-2n}/b, & q^{-2n}/c \end{bmatrix} ; q \Big]_n$$
$$\times \, {}_4\phi_3 \begin{bmatrix} q^{-n}, & b, & c, & q^{1-2n}/de \\ & q^{1-2n}/d, & q^{-2n}/e, & bcq^{n+1} \end{bmatrix} ; q \Big]$$

which is equivalent, in view of the Sears formula (2.2b), to the transformation established in Example 2.11.

c. For $(\delta; \alpha, \gamma) = (1; 0, 1)$, the corresponding transform reads as, based on the evaluation from Example 1.5,

$$\Omega_1 \begin{bmatrix} 0,1,1,0; & 2 \\ b,c,d,e; & n \end{bmatrix} = \begin{bmatrix} q^{-1-n}, cdq^{2n-1} \\ c/q, & d/q \end{bmatrix} \times \begin{bmatrix} q^{-1-2n}, & q^{1-2n}/bc \\ q^{-2n}/b, & q^{1-2n}/c \end{bmatrix} ; q \Big]_n$$
$$\times \, {}_4\phi_3 \begin{bmatrix} q^{-n}, & b, & c/q, & q^{1-2n}/de \\ & q^{1-2n}/d, & q^{-2n}/e, & bcq^n \end{bmatrix} ; q \Big]$$

which is equivalent, in view of the Sears formula (2.2b), to the transformation established in Example 2.13.

d. For $(\delta; \alpha, \gamma) = (1; 0, -1)$, the corresponding transform reads as, based on the evaluation from Example 1.6,

$$\Omega_1 \begin{bmatrix} 0,-1,1,0; & 1 \\ b, & c, & d,e; & n \end{bmatrix} = \begin{bmatrix} q^{1+n}, & qc/d \\ cq^{1+n}, & q/d \end{bmatrix} \times \begin{bmatrix} q^{-1-2n}, & q^{-2n}/bc \\ q^{-2n}/b, & q^{-1-2n}/c \end{bmatrix} ; q \Big]_n$$
$$\times \, {}_4\phi_3 \begin{bmatrix} q^{-n}, & b, & c, & q^{1-2n}/de \\ & q^{1-2n}/d, & q^{-2n}/e, & bcq^{n+1} \end{bmatrix} ; q \Big]$$

which is equivalent, in view of the Sears formula (2.2b), to the transformation established in Example 2.10.

2.5 Case ($\beta = \rho = 2 - \delta$)

The transforms and evaluations listed here are the reversals of those demonstrated in Section 2.2.

Example 2.31 *The reversals of the transforms from* Example 2.19 *may be stated as*

$$\Omega_0 \begin{bmatrix} 2,2,2,2; & \varepsilon \\ b,c,d,e; & n \end{bmatrix} \stackrel{\varepsilon=4,5}{=\!=\!=\!=} q^{n(2\varepsilon-11)} \begin{bmatrix} q^{-2n}, & q^{2-2n}/bc \\ q^{2-2n}/b, & q^{2-2n}/c \end{bmatrix} ; q \end{bmatrix}_n$$

$$\times \begin{bmatrix} bq^{n-1}, & cq^{n-1}, & dq^{2n-1}, & eq^{2n-1} \\ b/q, & c/q, & d/q, & e/q \end{bmatrix}$$

$$\times \; {}_4\phi_3 \begin{bmatrix} q^{-n}, & b/q, & c/q, & q^{2-2n}/de \\ & q^{1-2n}/d, q^{1-2n}/e, & bcq^{n-1} \end{bmatrix} ; q \end{bmatrix}$$

which reduce, by means of the Saalschütz formula (1.1) for $bcde = q^{3-3n}$, to the evaluations

$$ {}_5\phi_4 \begin{bmatrix} q^{-2n}, & b, & c, & d, & e \\ & q^{2-2n}/b, & q^{2-2n}/c, & q^{2-2n}/d, & q^{2-2n}/e \end{bmatrix} ; q^\varepsilon \end{bmatrix}$$

$$\stackrel{\varepsilon=1,2}{=\!=\!=\!=} q^{n(2\varepsilon-3)} \begin{bmatrix} bq^{n-1}, & cq^{n-1}, & dq^{n-1}, & bcdq^{2n-2} \\ b/q, & c/q, & d/q, & bcdq^{3n-2} \end{bmatrix}$$

$$\times \begin{bmatrix} q^{-2n}, & q^{2-2n}/bc, & q^{2-2n}/bd, & q^{2-2n}/cd \\ q^{2-2n}/b, & q^{2-2n}/c, & q^{2-2n}/d, & q^{2-2n}/bcd \end{bmatrix} ; q \end{bmatrix}_n .$$

Example 2.32 *The reversals of the transform from* Example 2.21 *may be stated as*

$$\Omega_0 \begin{bmatrix} 2,3,2,2; & 5 \\ b,c,d,e; & n \end{bmatrix} = q^{-3n} \begin{bmatrix} q^{-2n}, & q^{3-2n}/bc \\ q^{2-2n}/b, & q^{3-2n}/c \end{bmatrix} ; q \end{bmatrix}_n$$

$$\times \begin{bmatrix} bq^{n-1}, cq^{2n-1}, cq^{n-2}, dq^{2n-1}, eq^{2n-1} \\ b/q, & c/q, & c/q^2, & d/q, & e/q \end{bmatrix}$$

$$\times \, {}_4\phi_3 \begin{bmatrix} q^{-n}, & b/q, & c/q^2, & q^{2-2n}/de \\ & q^{1-2n}/d, q^{1-2n}/e, & bcq^{n-2} \end{bmatrix} ; q \end{bmatrix}$$

which reduces, by means of the Saalschütz formula (1.1) for $bcde = q^{4-3n}$, to the evaluation

$${}_5\phi_4 \begin{bmatrix} q^{-2n}, & b, & c, & d, & e \\ & q^{2-2n}/b, & q^{3-2n}/c, & q^{2-2n}/d, & q^{2-2n}/e \end{bmatrix} ; q \end{bmatrix}$$

$$= q^{-n} \begin{bmatrix} bq^{n-1}, & cq^{2n-1}, & cq^{n-2} & dq^{n-1}, & bcdq^{2n-3} \\ b/q, & c/q, & c/q^2, & d/q, & bcdq^{3n-3} \end{bmatrix}$$

$$\times \begin{bmatrix} q^{-2n}, & q^{3-2n}/bc, & q^{2-2n}/bd, & q^{3-2n}/cd \\ q^{2-2n}/b, & q^{3-2n}/c, & q^{2-2n}/d, & q^{3-2n}/bcd \end{bmatrix} ; q \end{bmatrix}_n .$$

Example 2.33 *The reversals of the transform from Example 2.22 may be stated as*

$$\Omega_1 \begin{bmatrix} 1,1,1,1; & 3 \\ b,c,d,e; & n \end{bmatrix} = -q^{-2n} \begin{bmatrix} q^{-1-2n}, & q^{1-2n}/bc \\ q^{1-2n}/b, & q^{1-2n}/c \end{bmatrix}_{n+1}$$

$$\times \begin{bmatrix} bq^{n-1}, & cq^{n-1}, & dq^{2n}, & eq^{2n} \\ b/q, & c/q, & d/q, & e/q \end{bmatrix}$$

$$\times {}_4\phi_3 \begin{bmatrix} q^{-n-1}, & b/q, & c/q, & q^{1-2n}/de \\ & q^{-2n}/d, q^{-2n}/e, & bcq^{n-1} \end{bmatrix}$$

which reduces, by means of the Saalschütz formula (1.1) for $bcde = q^{2-3n}$, to the evaluation

$$ {}_5\phi_4 \begin{bmatrix} q^{-1-2n}, & b, & c, & d, & e \\ & q^{1-2n}/b, & q^{1-2n}/c, & q^{1-2n}/d, & q^{1-2n}/e \end{bmatrix}$$

$$= -q \begin{bmatrix} bq^{n-1}, & cq^{n-1}, & dq^{n-1}, & bcdq^{2n-1} \\ b/q, & c/q, & d/q, & bcdq^{3n-1} \end{bmatrix}$$

$$\times \begin{bmatrix} q^{-1-2n}, & q^{1-2n}/bc, & q^{1-2n}/bd, & q^{1-2n}/cd \\ q^{1-2n}/b, & q^{1-2n}/c, & q^{1-2n}/d, & q^{1-2n}/bcd \end{bmatrix}_{n+1}.$$

Example 2.34 *The reversal version of the transform from Example 2.23 may be stated as*

$$\Omega_1 \begin{bmatrix} 2,2,1,1; & 4 \\ b,c,d,e; & n \end{bmatrix} = -q^{1-2n} \begin{bmatrix} q^{-2-2n}, & q^{2-2n}/bc \\ q^{2-2n}/b, & q^{2-2n}/c \end{bmatrix}_{n+1}$$

$$\times \begin{bmatrix} bq^{n-1}, & bq^{n-2}, & cq^{n-1}, & cq^{n-2}, & dq^{2n}, & eq^{2n} \\ b/q, & b/q^2, & c/q, & c/q^2, & d/q, & e/q \end{bmatrix}$$

$$\times {}_5\phi_4 \begin{bmatrix} q^{-1-n}, q^{-1-2n}, & b/q^2, & c/q^2, & q^{1-2n}/de \\ & q^{-2-2n}, q^{-2n}/d, q^{-2n}/e, & bcq^{n-2} \end{bmatrix}$$

which reduces, by means of the 2-balanced separation (2.4) via substituting $w \to q^{-2-2n}$ for $bcde = q^{3-3n}$, to the evaluation

$$ {}_5\phi_4 \begin{bmatrix} q^{-1-2n}, & b, & c, & d, & e \\ & q^{2-2n}/b, q^{2-2n}/c, q^{1-2n}/d, q^{1-2n}/e \end{bmatrix}$$

$$= -q^2 \left\{ 1 + q^{n+1} \begin{bmatrix} d/q, & bcdq^{2n-3} \\ bdq^{2n-1}, & cdq^{2n-1} \end{bmatrix} \right\}$$

$$\times \begin{bmatrix} bq^{n-1}, & bq^{n-2}, & cq^{n-1}, & cq^{n-2}, & dq^{2n}, & bcdq^{n-3} \\ b/q, & b/q^2, & c/q, & c/q^2, & d/q, & bcdq^{3n-2} \end{bmatrix}$$

$$\times \begin{bmatrix} q^{-1-2n}, & q^{2-2n}/bc, & q^{1-2n}/bd, & q^{1-2n}/cd \\ q^{2-2n}/b, & q^{2-2n}/c, & q^{-2n}/d, & q^{3-2n}/bcd \end{bmatrix}_{n+1}.$$

Remark

a. The reversal of the transform from Example 2.20 may be stated as

$$\Omega_0 \begin{bmatrix} 2,1,2,2; & 4 \\ b,c,d,e; & n \end{bmatrix} = q^{-2n} \begin{bmatrix} bq^{n-1}, dq^{2n-1}, eq^{2n-1} \\ b/q, \quad d/q, \quad e/q \end{bmatrix} \begin{bmatrix} q^{-2n}, q^{2-2n}/bc \\ q^{2-2n}/b, q^{1-2n}/c \end{bmatrix}_n$$

$$\times \,_4\phi_3 \begin{bmatrix} q^{-n}, & b/q, & c/q, & q^{2-2n}/de \\ & q^{1-2n}/d, & q^{1-2n}/e, & bcq^{n-1} \end{bmatrix}$$

which is equivalent, in view of the Sears formula (2.2b), to the transformation established in Example 2.28.

b. The reversal of the transform remarked at the end of Section 2.2 may be stated as

$$\Omega_1 \begin{bmatrix} 0,0,1,1; & 2 \\ b,c,d,e; & n \end{bmatrix} = \begin{bmatrix} q^{-1-n}, deq^{2n-1} \\ d/q, \quad e/q \end{bmatrix} \times \begin{bmatrix} q^{-1-2n}, q^{-2n}/bc \\ q^{-2n}/b, q^{-2n}/c \end{bmatrix}_n$$

$$\times \,_4\phi_3 \begin{bmatrix} q^{-n}, & b, & c, & q^{2-2n}/de \\ & q^{1-2n}/d, & q^{1-2n}/e, & bcq^{n+1} \end{bmatrix}$$

which is equivalent, in view of the Hall formula (2.2c), to the transformation established in Example 2.13.

2.6 Exceptional cases

Here we demonstrate some transforms and the related evaluations which may not be derived from the previous sections.

Example 2.35 *Taking $\lambda = \mu = \nu = -2$, the general relation (2.3) may be specified as*

$$
\Omega_1 \begin{bmatrix} -2, & -2, & -2, & -2; & -3 \\ b, & c, & d, & e; & n \end{bmatrix}
$$

$$
= \sum_k \begin{bmatrix} b, & c, & q^{-2-2n}/de \\ q^{-2-2n}/b, & q^{-2-2n}/c, & q^{-2-2n}/d, & q^{-2-2n}/e \end{bmatrix} ; q \end{bmatrix}_k
$$

$$
\times \ (-1)^k \frac{(q^{-1-2n}; q)_k \, (q^{-3-2n}; q)_{2k}}{(q; q)_k \, (q^{-3-2n}; q)_k} q^{-k(1+n)-\binom{k}{2}} / (bc)^k
$$

$$
\times \ _4\phi_3 \begin{bmatrix} q^{-3-2n+2k}, q^{-1-2n+k}, & bq^k, & cq^k \\ & q^{-3-2n+k}, q^{-2-2n+k}/b, q^{-2-2n+k}/c \end{bmatrix} ; \frac{q^{-1-n-k}}{bc} \end{bmatrix}
$$

where the $_4\phi_3[\cdots]$ may be evaluated by Example 1.7 *through setting $u \to q^{k-2-2n}$ and $v \to q^{k-3-2n}$, as follows*

$$
_4\phi_3 \begin{bmatrix} q^{-3-2n+2k}, & q^{-1-2n+k}, & bq^k, & cq^k \\ & q^{-3-2n+k}, & q^{-2-2n+k}/b, & q^{-2-2n+k}/c \end{bmatrix} ; \frac{q^{-1-n-k}}{bc} \end{bmatrix}
$$

$$
= \frac{(1-q^{k-2-n})(1-q^{-2-2n})}{(1-q^{k-2-2n})(1-q^{k-3-2n})} \begin{bmatrix} q^{2k-3-2n}, & q^{-2-2n}/bc \\ q^{k-2-2n}/b, & q^{k-2-2n}/c \end{bmatrix} ; q \end{bmatrix}_{1+n-k}.
$$

The corresponding new transform reads as

$$
\Omega_1 \begin{bmatrix} -2, -2, -2, -2; & -3 \\ b, & c, & d, & e; & n \end{bmatrix} = \begin{bmatrix} q^{-2-2n}, & q^{-2-2n}/bc \\ q^{-2-2n}/b, & q^{-2-2n}/c \end{bmatrix} ; q \end{bmatrix}_{n+1}
$$

$$
\times \ _4\phi_3 \begin{bmatrix} q^{-1-n}, & b, & c, & q^{-2-2n}/de \\ & q^{-2-2n}/d, & q^{-2-2n}/e, & bcq^{n+2} \end{bmatrix} ; q \end{bmatrix}.
$$

When $bcde = q^{-4-3n}$, it reduces, by means of the Saalschütz formula (1.1), to the evaluation

$$
_5\phi_4 \begin{bmatrix} q^{-1-2n}, & b, & c, & d, & e \\ & q^{-2-2n}/b, & q^{-2-2n}/c, & q^{-2-2n}/d, & q^{-2-2n}/e \end{bmatrix} ; q \end{bmatrix}
$$

$$
= \begin{bmatrix} q^{-2-2n}, & q^{-2-2n}/bc, & q^{-2-2n}/bd, & q^{-2-2n}/cd \\ q^{-2-2n}/b, & q^{-2-2n}/c, & q^{-2-2n}/d, & q^{-2-2n}/bcd \end{bmatrix} ; q \end{bmatrix}_{n+1}.
$$

Example 2.36 *The reversal of the transform from* Example 2.35 *may be stated as*

$$\Omega_1 \begin{bmatrix} 2,2,2,2; & 5 \\ b,c,d,e; & n \end{bmatrix} = -q^{-4n} \begin{bmatrix} q^{-2-2n}, & q^{2-2n}/bc \\ q^{2-2n}/b, & q^{2-2n}/c \end{bmatrix} ; q \end{bmatrix}_{n+1}$$

$$\times \begin{bmatrix} bq^{n-1}, & bq^{n-2}, & cq^{n-1}, & cq^{n-2} \\ b/q, & b/q^2, & c/q, & c/q^2 \end{bmatrix}$$

$$\times \begin{bmatrix} dq^{2n}, & dq^{2n-1}, & eq^{2n}, & eq^{2n-1} \\ d/q, & d/q^2, & e/q, & e/q^2 \end{bmatrix}$$

$$\times {}_4\phi_3 \begin{bmatrix} q^{-1-n}, & b/q^2, & c/q^2, & q^{2-2n}/de \\ & q^{-2n}/d, & q^{-2n}/e, & bcq^{n-2} \end{bmatrix} ; q \end{bmatrix}$$

which reduces, by means of the Saalschütz formula (1.1) *for* $bcde = q^{4-3n}$, *to the evaluation*

$$_5\phi_4 \begin{bmatrix} q^{-1-2n}, & b, & c, & d, & e \\ & q^{2-2n}/b, & q^{2-2n}/c, & q^{2-2n}/d, & q^{2-2n}/e \end{bmatrix} ; q \end{bmatrix}$$

$$= -q^2 \begin{bmatrix} q^{-2-2n}, & q^{2-2n}/bc, & q^{2-2n}/bd, & q^{2-2n}/cd \\ q^{2-2n}/b, & q^{2-2n}/c, & q^{2-2n}/d, & q^{2-2n}/bcd \end{bmatrix} ; q \end{bmatrix}_{n+1}$$

$$\times \begin{bmatrix} bq^{n-1}, bq^{n-2}, cq^{n-1}, cq^{n-2}, dq^{n-1}, dq^{n-2}, bcdq^{2n-2}, bcdq^{2n-3} \\ b/q, & b/q^2, & c/q, & c/q^2, & d/q, & d/q^2, & bcdq^{3n-2}, bcdq^{3n-3} \end{bmatrix}.$$

Remark

a. Taking $\lambda = \nu = 0$ and $\mu = 1$, the general relation (2.3) may be specified as

$$\Omega_1 \begin{bmatrix} 0,0,1,-1; & 1 \\ b,c,d, & e; & n \end{bmatrix} = \sum_k \begin{bmatrix} b, & c, & d, & q^{1-2n}/de \\ q^{-2n}/b, & q^{-2n}/c, & d/q, & q^{1-2n}/d, & q^{-1-2n}/e \end{bmatrix} ; q \end{bmatrix}_k$$

$$\times \qquad (-1)^k \frac{(q^{-1-2n}; q)_{2k}}{(q; q)_k} q^{-kn-\binom{k}{2}} / (bc)^k$$

$$\times {}_5\phi_4 \begin{bmatrix} q^{-1-2n+2k}, q^k d, & q^{-2n+k}/e, & bq^k, & cq^k \\ & q^{k-1}d, q^{-1-2n+k}/e, q^{-2n+k}/b, q^{-2n+k}/c \end{bmatrix} ; \frac{q^{-n-k}}{bc} \end{bmatrix}$$

where the $_5\phi_4[\cdots]$ *may be evaluated by Example 1.7 through setting* $u \to dq^{k-1}$ *and* $v \to q^{-1-2n+k}/e$. *The corresponding transform reads as*

$$\Omega_1 \begin{bmatrix} 0,0,1,-1; & 1 \\ b,c,d, & e; & n \end{bmatrix} = q^n \begin{bmatrix} q^{n+1}, & qe/d \\ eq^{2n+1}, & q/d \end{bmatrix} \times \begin{bmatrix} q^{-1-2n}, q^{-2n}/bc \\ q^{-2n}/b, & q^{-2n}/c \end{bmatrix} ; q \end{bmatrix}_n$$

$$\times {}_4\phi_3 \begin{bmatrix} q^{-n}, & b, & c, & q^{1-2n}/de \\ & q^{1-2n}/d, & q^{-2n}/e, & bcq^{n+1} \end{bmatrix} ; q \end{bmatrix}$$

which is equivalent, in view of the Hall formula (2.2c), *to the transformation established in* Example 2.10.

b. Taking $\lambda = \mu = 0$ and $\nu = 2$, the general relation (2.3) may be specified as

$$\Omega_1 \begin{bmatrix} 0,0,0,2; & 2 \\ b,c,d,e; & n \end{bmatrix} = \sum_k \begin{bmatrix} b, & c, & e, & q^{2-2n}/de \\ q^{-2n}/b, & q^{-2n}/c, & e/q^2, & q^{-2n}/d, & q^{2-2n}/e \end{bmatrix}_k ; q$$

$$\times \quad (-1)^k \frac{(q^{-1-2n}; q)_{2k}}{(q; q)_k} q^{-kn-\binom{k}{2}} / (bc)^k$$

$$\times {}_4\phi_3 \begin{bmatrix} q^{-1-2n+2k}, & eq^k, & bq^k, & cq^k \\ eq^{k-2}, q^{-2n+k}/b, q^{-2n+k}/c & ; & \frac{q^{-n-k}}{bc} \end{bmatrix}$$

where the ${}_4\phi_3[\cdots]$ may be evaluated by Example 1.7 through setting $u \to eq^{k-1}$ and $v \to eq^{k-2}$. The corresponding transform reads as

$$\Omega_1 \begin{bmatrix} 0,0,0,2; & 2 \\ b,c,d,e; & n \end{bmatrix} = \begin{bmatrix} q^{-n-1}, e^2 q^{2n-2} \\ e/q, & e/q^2 \end{bmatrix} \times \begin{bmatrix} q^{-1-2n}, q^{-2n}/bc \\ q^{-2n}/b, & q^{-2n}/c \end{bmatrix}_n ; q$$

$$\times {}_4\phi_3 \begin{bmatrix} q^{-n}, & b, & c, & q^{2-2n}/de \\ & q^{-2n}/d, & q^{2-2n}/e, & bcq^{n+1} \end{bmatrix} ; q$$

which is equivalent, in view of the Sears formula (2.2b), to the transformation established in Example 2.15.

c. Taking $\lambda = \mu = \nu = 0$, the general relation (2.3) may be specified as

$$\Omega_1 \begin{bmatrix} 0,0,0,-2; & 0 \\ b,c,d, & e; & n \end{bmatrix} = \sum_k \begin{bmatrix} b, & c, & q^{-2n}/de \\ q^{-2n}/b, & q^{-2n}/c, & q^{-2n}/d, & q^{-2-2n}/e \end{bmatrix}_k ; q$$

$$\times \quad (-1)^k \frac{(q^{-1-2n}; q)_{2k}}{(q; q)_k} q^{-kn-\binom{k}{2}} / (bc)^k$$

$$\times {}_4\phi_3 \begin{bmatrix} q^{-1-2n+2k}, q^{-2n+k}/e, & bq^k, & cq^k \\ q^{-2-2n+k}/e, q^{-2n+k}/b, q^{-2n+k}/c & ; & \frac{q^{-n-k}}{bc} \end{bmatrix}$$

where the ${}_4\phi_3[\cdots]$ may be evaluated by Example 1.7 through setting $u \to q^{-1-2n+k}/e$ and $v \to q^{-2-2n+k}/e$. The corresponding transform reads as

$$\Omega_1 \begin{bmatrix} 0,0,0,-2; & 0 \\ b,c,d, & e; & n \end{bmatrix} = \begin{bmatrix} q^{-1-n}, q^{-2-2n}/e^2 \\ q^{-1-2n}/e, q^{-2-2n}/e \end{bmatrix} \times \begin{bmatrix} q^{-1-2n}, q^{-2n}/bc \\ q^{-2n}/b, q^{-2n}/c \end{bmatrix}_n ; q$$

$$\times {}_4\phi_3 \begin{bmatrix} q^{-n}, & b, & c, & q^{-2n}/de \\ & q^{-2n}/d, & q^{-2n}/e, & bcq^{n+1} \end{bmatrix} ; q$$

which is equivalent, in view of the Sears formula (2.2b), to the transformation established in Example 2.16.

2.7   Equivalent classes

Similar to Section 1.5, we can classify the transforms for $_5\phi_4[\cdots]$.

It is easy to verify that the inversion operator $\mathcal{C}'$ introduced in Section 1.5 converts an almost poised $_5\phi_4[\cdots]$ to a series of the same kind. As an extension of (1.18), the reversal operation yields, through (1.17b), the relation

$$\Omega_\delta \begin{bmatrix} \alpha, \gamma, \beta, \rho;\ \varepsilon \\ b,\ c,\ d,\ e;\ n \end{bmatrix} = \quad (-1)^\delta \quad q^{(\delta+2n)\,(\varepsilon-\alpha-\gamma-\beta-\rho+\frac{3-5\delta}{2})} \tag{2.10}$$

$$\times \begin{bmatrix} b, & c, & d, & e \\ ba^{1-\alpha-\delta}, cq^{1-\gamma-\delta}, dq^{1-\beta-\delta}, eq^{1-\rho-\delta} \end{bmatrix} ;\ q \Big]_{\delta+2n}$$

$$\times \Omega_\delta \begin{bmatrix} 2-\alpha-2\delta,\, 2-\gamma-2\delta,\, 2-\beta-2\delta,\, 2-\rho-2\delta;\ 5-\varepsilon-3\delta \\ bq^{1-\alpha-\delta},\, cq^{1-\gamma-\delta},\, dq^{1-\beta-\delta},\, eq^{1-\rho-\delta};\qquad n \end{bmatrix}.$$

Therefore we may define two almost-poised transforms by

$$\mathcal{C}^* : \Omega_\delta \begin{bmatrix} \alpha, \gamma, \beta, \rho;\ \varepsilon \\ b,\ c,\ d,\ e;\ n \end{bmatrix} \longmapsto \Omega_\delta \begin{bmatrix} \alpha, \gamma, \beta, \rho;\, 1-\varepsilon+\delta+\alpha+\gamma+\beta+\rho \\ b,\ c,\ d,\ e;\qquad\qquad n \end{bmatrix} \tag{2.11a}$$

$$\mathcal{R}^* : \Omega_\delta \begin{bmatrix} \alpha, \gamma, \beta, \rho;\ \varepsilon \\ b,\ c,\ d,\ e;\ n \end{bmatrix} \longmapsto \Omega_\delta \begin{bmatrix} 2-\alpha-2\delta,\, 2-\gamma-2\delta,\, 2-\beta-2\delta, \\ b, & c, & d, \\ & 2-\rho-2\delta;\ 5-\varepsilon-3\delta \\ & e;\qquad\qquad n \end{bmatrix}. \tag{2.11b}$$

They generate a Coxeter group with 4-elements $\mathbf{G}^* = \{\mathcal{I}^*, \mathcal{C}^*, \mathcal{R}^*, \mathcal{C}^*\mathcal{R}^* = \mathcal{R}^*\mathcal{C}^*\}$, where $\mathcal{I}^*$ is an identity and

$$\mathcal{C}^*\mathcal{R}^* : \Omega_\delta \begin{bmatrix} \alpha, \gamma, \beta, \rho;\ \varepsilon \\ b,\ c,\ d,\ e;\ n \end{bmatrix} \longmapsto \Omega_\delta \begin{bmatrix} 2-\alpha-2\delta,\, 2-\gamma-2\delta,\, 2-\beta-2\delta, \\ b, & c, & d, \\ 2-\rho-2\delta;\ 4+\varepsilon-4\delta-\alpha-\gamma-\beta-\rho \\ e;\qquad\qquad n \end{bmatrix}. \tag{2.11c}$$

Under the action of this group $\mathbf{G}^*$, we can classify, according to its orbits, the almost poised transforms demonstrated in the previous examples.

\multicolumn{3}{c}{Equivalent classes for $\Omega_0 \begin{bmatrix} \alpha, \gamma, \beta, \rho; & \varepsilon \\ b,\ c,\ d,\ e; & n \end{bmatrix}$}

Nr	$\mathbf{G}^*$-orbits: $\{(\alpha,\ \gamma,\ \beta,\ \rho;\ \varepsilon)\}$	Note
1	(1,1,1,1;2), (1,1,1,1;3)	Ex. 2.1
2	(2,1,1,0;2), (2,1,1,0;3)	Ex. 2.2
3	(2,1,1,1;3), (1,1,1,0;2)	Exs. 2.3, 2.4
4	(2,2,1,0;3), (2,1,0,0;2)	Exs. 2.24, 2.27
5	(2,2,1,1;3), (1,1,0,0;1), (2,2,1,1;4), (1,1,0,0;2)	Exs. 2.5, 2.6
6	(2,2,2,1;4), (1,0,0,0;1)	Exs. 2.20, 2.28
7	(2,2,2,2;4), (0,0,0,0;0), (2,2,2,2;5), (0,0,0,0;1)	Exs. 2.19, 2.31
8	(3,2,1,1;4), (0,1,1,-1;1)	Exs. 2.7, 2.8
9	(3,2,2,2;5), (0,0,0,-1;0)	Exs. 2.21, 2.32

Equivalent classes for $\Omega_1 \begin{bmatrix} \alpha, \gamma, \beta, \rho; & \varepsilon \\ b,\ c,\ d,\ e; & n \end{bmatrix}$		
Nr	$\mathbf{G}^*$-orbits: $\{(\alpha,\ \gamma,\ \beta,\ \rho;\ \varepsilon)\}$	Note
10	(0,0,0,0;0), (0,0,0,0;1), (0,0,0,0;2)	Ex. 2.9
11	(1,-1,0,0;1)	Ex. 2.10
12	(1,0,0,0;1), (0,0,0,-1;0), (1,0,0,0;2), (0,0,0,-1;1)	Exs. 2.11, 2.12
13	(1,1,0,0;2), (0,0,-1,-1;0)	Exs. 2.13, 2.14
14	(2,0,0,0;2), (0,0,0,-2;0)	Exs. 2.15, 2.16
15	(1,1,1,0;2), (0,-1,-1,-1;0), (1,1,1,0;3), (0,-1,-1,-1;-1)	Exs. 2.25, 2.29
16	(1,1,1,1;3), (-1,-1,-1,-1;-1)	Exs. 2.22, 2.33
17	(2,1,1,0;3), (0,-1,-1,-2;-1)	Exs. 2.26, 2.30
18	(2,2,0,0;3), (0,0,-2,-2;-1)	Exs. 2.17, 2.18
19	(2,2,1,1;4), (-1,-1,-2,-2;-2)	Exs. 2.23, 2.34
20	(2,2,2,2;5), (-2,-2,-2,-2;-3)	Exs. 2.35, 2.36

3 Bilateral Almost-Poised Evaluations

Recalling the bilateral basic hypergeometric series defined by (0.3), we have two important relations

$$_r\psi_s \begin{bmatrix} a_1, a_2, \cdots, a_r \\ b_1, b_2, \cdots, b_s \end{bmatrix} = \sum_{n=-\infty}^{+\infty} \begin{bmatrix} q/b_1, q/b_2, \cdots, q/b_s \\ q/a_1, q/a_2, \cdots, q/a_r \end{bmatrix}_n \{\frac{b_1 b_2 \cdots b_s}{z a_1 a_2 \cdots a_r}\}^n \quad (3.1a)$$

$$_r\psi_s \begin{bmatrix} a_1, a_2, \cdots, a_r \\ b_1, b_2, \cdots, b_s \end{bmatrix} = \{(-1)^n q^{\binom{n}{2}}\}^{s-r} \begin{bmatrix} a_1, a_2, \cdots, a_r \\ b_1, b_2, \cdots, b_s \end{bmatrix}_n z^n \quad (3.1b)$$

$$\times {}_r\psi_s \begin{bmatrix} a_1 q^n, a_2 q^n, \cdots, a_r q^n \\ b_1 q^n, b_2 q^n, \cdots, b_s q^n \end{bmatrix} zq^{n(s-r)} \end{bmatrix} \quad (3.1c)$$

by reversing and shifting the summation index,

In the way similar to Bailey's [3], one can get two bilateral basic hypergeometric formulae.

Taking $a \to 1$ in the Watson transformation (2.1a-2.1b) and using that

$$(x)_k = (-1)^k q^{\binom{k}{2}} x^k / (q/x)_{-k}$$

we can rewrite

$$[Eq.\ 2.1a] \Rightarrow 1 + \sum_{k=1}^{n} (1+q^k) \begin{bmatrix} q^{-n}, & b, & c, & d, & e \\ q^{1+n}, q/b, q/c, q/d, q/e \end{bmatrix}_k (\frac{q^{2+n}}{bcde})^k$$

$$= \sum_{k=0}^{n} \begin{bmatrix} q^{-n}, & b, & c, & d, & e \\ q^{1+n}, q/b, q/c, q/d, q/e \end{bmatrix}_k (\frac{q^{2+n}}{bcde})^k$$

$$+ \sum_{k=1}^{n} \begin{bmatrix} q^{-n}, & b, & c, & d, & e \\ q^{1+n}, q/b, q/c, q/d, q/e \end{bmatrix}_k (\frac{q^{3+n}}{bcde})^k$$

$$= \sum_{k=0}^{n} \begin{bmatrix} q^{-n}, & b, & c, & d, & e \\ q^{1+n}, q/b, q/c, q/d, q/e \end{bmatrix}_k (\frac{q^{2+n}}{bcde})^k$$

$$+ \sum_{k=-n}^{-1} \begin{bmatrix} q^{-n}, & b, & c, & d, & e \\ q^{1+n}, q/b, q/c, q/d, q/e \end{bmatrix}_k (\frac{q^{2+n}}{bcde})^k$$

$$= {}_5\psi_5 \begin{bmatrix} q^{-n}, & b, & c, & d, & e \\ q^{1+n}, & q/b, & q/c, & q/d, & q/e \end{bmatrix}; \frac{q^{2+n}}{bcde} \end{bmatrix}.$$

61

Similarly, taking $a \to q$ instead, in the Watson transformation (2.1a-2.1b) we can rewrite

$$(1-q)[Eq.\ 2.1a] \quad \Rightarrow \quad \sum_{k=0}^{n}(1-q^{1+2k}) \left[\begin{matrix} q^{-n}, & b, & c, & d, & e \\ q^{2+n}, & q^2/b, & q^2/c, & q^2/d, & q^2/e \end{matrix} ; q\right]_k \left(\frac{q^{4+n}}{bcde}\right)^k$$

$$= \sum_{k=0}^{n} \left[\begin{matrix} q^{-n}, & b, & c, & d, & e \\ q^{2+n}, & q^2/b, & q^2/c, & q^2/d, & q^2/e \end{matrix} ; q\right]_k \left(\frac{q^{4+n}}{bcde}\right)^k$$

$$- q\sum_{k=0}^{n} \left[\begin{matrix} q^{-n}, & b, & c, & d, & e \\ q^{2+n}, & q^2/b, & q^2/c, & q^2/d, & q^2/e \end{matrix} ; q\right]_k \left(\frac{q^{6+n}}{bcde}\right)^k$$

$$= \sum_{k=0}^{n} \left[\begin{matrix} q^{-n}, & b, & c, & d, & e \\ q^{2+n}, & q^2/b, & q^2/c, & q^2/d, & q^2/e \end{matrix} ; q\right]_k \left(\frac{q^{4+n}}{bcde}\right)^k$$

$$+ \sum_{k=-1-n}^{-1} \left[\begin{matrix} q^{-n}, & b, & c, & d, & e \\ q^{2+n}, & q^2/b, & q^2/c, & q^2/d, & q^2/e \end{matrix} ; q\right]_k \left(\frac{q^{4+n}}{bcde}\right)^k$$

$$= \quad _5\psi_5 \left[\begin{matrix} q^{-n}, & b, & c, & d, & e \\ q^{2+n}, & q^2/b, & q^2/c, & q^2/d, & q^2/e \end{matrix} ; \frac{q^{4+n}}{bcde}\right]$$

$$= \quad -q\, _5\psi_5 \left[\begin{matrix} q^{-1-n}, & b/q, & c/q, & d/q, & e/q \\ q^{1+n}, & q/b, & q/c, & q/d, & q/e \end{matrix} ; \frac{q^{4+n}}{bcde}\right]$$

where the last line follows from the previous one under shifting the summation index by one.

From [Eq. 2.1a]=[Eq. 2.1b] with parameter a specified as above, we have

$$_5\psi_5 \left[\begin{matrix} q^{-n}, & b, & c, & d, & e \\ q^{1+n}, & q/b, & q/c, & q/d, & q/e \end{matrix} ; \frac{q^{2+n}}{bcde}\right] \tag{3.2a}$$

$$= \left[\begin{matrix} q, & q/bc \\ q/b, & q/c \end{matrix} ; q\right]_n \times {}_4\phi_3 \left[\begin{matrix} q^{-n}, & b, & c, & q/de \\ & q/d, & q/e, & bcq^{-n} \end{matrix} ; q\right] \tag{3.2b}$$

$$_5\psi_5 \left[\begin{matrix} q^{-n}, & b, & c, & d, & e \\ q^{n}, & 1/b, & 1/c, & 1/d, & 1/e \end{matrix} ; \frac{q^{n-1}}{bcde}\right] \tag{3.2c}$$

$$= \frac{(1-b)\,(1-c)}{1-bcq} \left[\begin{matrix} q, & q^{-1}/bc \\ 1/b, & 1/c \end{matrix} ; q\right]_n \times {}_4\phi_3 \left[\begin{matrix} q^{1-n}, & bq, & cq, & 1/de \\ bcq^{2-n}, & q/d, & q/e \end{matrix} ; q\right] \tag{3.2d}$$

where the parameter replacements $n \to n-1$, $b \to bq$, $c \to cq$, $d \to dq$, and $e \to eq$ have been performed for the last trasformation.

Their limiting versions of $n \to \infty$ may be stated as transformations of bilateral series into unilateral series:

$$_4\psi_5 \left[\begin{array}{ccccc} & b, & c, & d, & e \\ 0, & q/b, & q/c, & q/d, & q/e \end{array} ; \frac{q^2}{bcde} \right] \tag{3.3a}$$

$$= \left[\begin{array}{c} q, q/bc \\ q/b, q/c \end{array} ; q \right]_\infty \times {}_3\phi_2 \left[\begin{array}{c} b, c, q/de \\ q/d, q/e \end{array} ; q/bc \right] \tag{3.3b}$$

$$_4\psi_5 \left[\begin{array}{ccccc} & b, & c, & d, & e \\ 0, & 1/b, & 1/c, & 1/d, & 1/e \end{array} ; \frac{q^{-1}}{bcde} \right] \tag{3.3c}$$

$$= \frac{-1}{q} \left[\begin{array}{c} q, 1/bc \\ q/b, q/c \end{array} ; q \right]_\infty \times {}_3\phi_2 \left[\begin{array}{c} bq, cq, 1/de \\ q/d, q/e \end{array} ; 1/bc \right]. \tag{3.3d}$$

The symmetric properties of these transforms on parameters b, c, d and e reveal some particular relations of ${}_3\phi_2$, known as the q-Kummer-Thomae-Whipple theorem and the Hall (1936) formula (cf. [11, Appendix III]):

$$_3\phi_2 \left[\begin{array}{c} a, b, c \\ d, e \end{array} ; \frac{de}{abc} \right] = \left[\begin{array}{c} e/a, de/bc \\ e, de/abc \end{array} ; q \right]_\infty {}_3\phi_2 \left[\begin{array}{c} a, d/b, d/c \\ d, de/bc \end{array} ; e/a \right] \tag{3.4a}$$

$$= \left[\begin{array}{c} a, de/ab, de/ac \\ d, e, de/abc \end{array} ; q \right]_\infty {}_3\phi_2 \left[\begin{array}{c} d/a, e/a, de/abc \\ de/ab, de/ac \end{array} ; a \right] \tag{3.4b}$$

When $e \to 0$, the transforms stated in (3.3) reduce to Bailey's [3] formulae:

$$_3\psi_3 \left[\begin{array}{ccc} b, & c, & d \\ q/b, & q/c, & q/d \end{array} ; \frac{q}{bcd} \right] = \left[\begin{array}{c} q, q/bc, q/bd, q/cd \\ q/b, q/c, q/d, q/bcd \end{array} ; q \right]_\infty \tag{3.5a}$$

$$_3\psi_3 \left[\begin{array}{ccc} b, & c, & d \\ 1/b, & 1/c, & 1/d \end{array} ; \frac{q^{-1}}{bcd} \right] = \frac{-1}{q} \left[\begin{array}{c} q, 1/bc, 1/bd, 1/cd \\ q/b, q/c, q/d, 1/bcdq \end{array} ; q \right]_\infty \tag{3.5b}$$

Now for basic almost poised series, we can restate it, by shifting the summation index, as follows:

$$\Omega_\delta \left[\begin{array}{c} \alpha, \gamma, \beta, \rho; \varepsilon \\ b, c, d, e; n \end{array} \right] = \left(\frac{q^{\varepsilon - 3n}}{bcde} \right)^n \left[\begin{array}{ccccc} q^{-\delta - 2n}, & b, & c, & d, & e \\ q, & q^{\alpha - 2n}/b, & q^{\gamma - 2n}/c, & q^{\beta - 2n}/d, & q^{\rho - 2n}/e \end{array} ; q \right]_n$$

$$\times {}_5\psi_5 \left[\begin{array}{ccccc} q^{-\delta - n}, & bq^n, & cq^n, & dq^n, & eq^n \\ q^{1+n}, & q^{\alpha - n}/b, & q^{\gamma - n}/c, & q^{\beta - n}/d, & q^{\rho - n}/e \end{array} ; \frac{q^{\varepsilon - 3n}}{bcde} \right].$$

Replacing b, c, d and e by bq^{-n}, cq^{-n}, dq^{-n} and eq^{-n} respectively, we find, after some trivial modification, that

$$_5\psi_5 \left[\begin{array}{ccccc} q^{-\delta - n}, & b, & c, & d, & e \\ q^{1+n}, & q^\alpha/b, & q^\gamma/c, & q^\beta/d, & q^\rho/e \end{array} ; \frac{q^{\varepsilon + n}}{bcde} \right] \tag{3.6a}$$

$$= \left(\frac{q^{\alpha + \gamma + \beta + \rho - \varepsilon - n}}{bcde} \right)^n \left[\begin{array}{ccccc} q, & bq^{1-\alpha}, cq^{1-\gamma}, dq^{1-\beta}, eq^{1-\rho} \\ q^{-\delta - 2n}, & q/b, & q/c, & q/d, & q/e \end{array} ; q \right]_n \tag{3.6b}$$

$$\times \quad \Omega_\delta \left[\begin{array}{ccccc} \alpha, & \gamma, & \beta, & \rho; & \varepsilon \\ bq^{-n}, & cq^{-n}, & dq^{-n}, & eq^{-n}; & n \end{array} \right]. \tag{3.6c}$$

For convenience, we will use the following notation to denote the bilateral basic almost poised series

$$\Theta\left[\begin{matrix} \alpha, & \beta, & \gamma \\ b, & c, & d \end{matrix}; \varepsilon\right] = {}_3\psi_3\left[\begin{matrix} b, & c, & d \\ q^\alpha/b, & q^\beta/c, & q^\gamma/d \end{matrix}; q^\varepsilon/bcd\right] \tag{3.7a}$$

with reversing and shifting operations

$$\Theta\left[\begin{matrix} \alpha, & \beta, & \gamma \\ b, & c, & d \end{matrix}; \varepsilon\right] \overset{\mathcal{R}}{=\!=} \Theta\left[\begin{matrix} 2-\alpha, & 2-\beta, & 2-\gamma \\ bq^{1-\alpha}, & cq^{1-\beta}, & dq^{1-\gamma} \end{matrix}; \; 3-\varepsilon\right] \tag{3.7b}$$

$$\Theta\left[\begin{matrix} \alpha, & \beta, & \gamma \\ b, & c, & d \end{matrix}; \varepsilon\right] \overset{\mathcal{S}^n}{=\!=} \left(\frac{q^\varepsilon}{bcd}\right)^n \left[\begin{matrix} b, & c, & d \\ q^\alpha/b, & q^\beta/c, & q^\gamma/d \end{matrix}; q\right]_n \tag{3.7c}$$

$$\times \; \Theta\left[\begin{matrix} \alpha+2n, & \beta+2n, & \gamma+2n \\ bq^n, & cq^n, & dq^n \end{matrix}; \varepsilon+3n\right]$$

where $\mathcal{R}$ is invariant under $\mathcal{S}$-conjugation:

$$\Theta\left[\begin{matrix} \alpha, & \beta, & \gamma \\ b, & c, & d \end{matrix}; \varepsilon\right] \overset{\mathcal{S}^n\mathcal{R}=\mathcal{R}\mathcal{S}^{-n}}{=\!=\!=\!=\!=\!=} \left(\frac{q^{\alpha+\beta+\gamma-\varepsilon}}{bcd}\right)^n \left[\begin{matrix} bq^{1-\alpha}, cq^{1-\beta}, dq^{1-\gamma} \\ q/b, \quad q/c, \quad q/d \end{matrix}; q\right]_n \tag{3.7d}$$

$$\times \Theta\left[\begin{matrix} 2-\alpha+2n, 2-\beta+2n, 2-\gamma+2n \\ bq^{1-\alpha+n}, \quad cq^{1-\beta+n}, \quad dq^{1-\gamma+n} \end{matrix}; \; 3-\varepsilon+3n\right]. \tag{3.7e}$$

Substituting the transformations of the Whipple type and evaluations established in Section 2 into (3.6c) and using the q-Gauss theorem

$$_2\phi_1\left[\begin{matrix} u, & v \\ & w \end{matrix}; w/uv\right] = \left[\begin{matrix} w/u, & w/v \\ w, & w/uv \end{matrix}; q\right]_\infty$$

we demonstrate a number of bilateral almost basic hypergeometric transforms and identities through the following examples.

Example 3.1 *Rewriting the transformations from* Example 2.9

$$\Omega_1\left[\begin{matrix} 0, & 0, & 0, & 0; & \varepsilon \\ bq^{-n}, & cq^{-n}, & dq^{-n}, & eq^{-n}; & n \end{matrix}\right] \quad (\varepsilon = 0, 1, 2)$$

$$= \{1-q^{(\varepsilon-1)(2n+1)}\}\{bcq^{1+n}\}^n \left[\begin{matrix} q^{-2n}, & 1/bc \\ bq, & cq \end{matrix}; q\right]_n$$

$$\times \quad _4\phi_3\left[\begin{matrix} q^{-n}, & bq^{-n}, & cq^{-n}, & 1/de \\ & q^{-n}/d, & q^{-n}/e, & q^{1-n}bc \end{matrix}; q\right]$$

with the above displayed $_4\phi_3$ *being replaced by its reversal*

$$_4\phi_3\left[\begin{matrix} q^{-n}, bq^{-n}, cq^{-n}, 1/de \\ q^{-n}/d, q^{-n}/e, q^{1-n}bc \end{matrix}; q\right] = (de)^n \left[\begin{matrix} q/b, & q/c, & 1/de \\ dq, & eq, & 1/bc \end{matrix}; q\right]_n$$

$$\times \quad _4\phi_3\left[\begin{matrix} q^{-n}, dq, eq, 1/bc \\ q/b, q/c, q^{1-n}de \end{matrix}; q\right]$$

we can specify (3.6) as three terminating series-transforms

$$
{}_5\psi_5\left[\begin{array}{ccccc} q^{-1-n}, & b, & c, & d, & e \\ q^{1+n}, & 1/b, & 1/c, & 1/d, & 1/e \end{array}; \frac{q^{\varepsilon+n}}{bcde}\right] \tag{3.8a}
$$

$$
\underset{\varepsilon=0,1,2}{=\!=\!=\!=}(1-q^{\varepsilon-1})\left[\begin{array}{c} q^2, 1/de \\ q/d, q/e \end{array}; q\right]_n {}_4\phi_3\left[\begin{array}{c} q^{-n}, dq, eq, 1/bc \\ q/b, q/c, q^{1-n}de \end{array}; q\right] \tag{3.8b}
$$

whose limiting versions read as

$$
{}_4\psi_5\left[\begin{array}{ccccc} & b, & c, & d, & e \\ 0, & 1/b, & 1/c, & 1/d, & 1/e \end{array}; \frac{q^\varepsilon}{bcde}\right] \quad (\varepsilon=-1,0,1) \tag{3.9a}
$$

$$
= (1-q^\varepsilon)\left[\begin{array}{c} q^2, 1/de \\ q/d, q/e \end{array}; q\right]_\infty {}_3\phi_2\left[\begin{array}{ccc} 1/bc, & dq, & eq \\ & q/b, & q/c \end{array}; \frac{1}{de}\right] \tag{3.9b}
$$

$$
= (1-q^\varepsilon)\left[\begin{array}{c} q^2, 1/bc \\ q/b, q/c \end{array}; q\right]_\infty {}_3\phi_2\left[\begin{array}{ccc} 1/de, & bq, & cq \\ & q/d, & q/e \end{array}; \frac{1}{bc}\right] \tag{3.9c}
$$

where the last line follows from the symmetry on parameters b, c, d and e, which is also a direct consequence of Hall's formula (3.4b).

When $e = 0$, the ${}_3\phi_2$ reduces to ${}_2\phi_1$ which may be evaluated by the q-Gauss theorem. In this case, we have three bilateral basic hypergeometric identities

$$
{}_3\psi_3\left[\begin{array}{ccc} b, & c, & d \\ 1/b, & 1/c, & 1/d \end{array}; \frac{q^\varepsilon}{bcd}\right] \quad (\varepsilon=-1,0,1) \tag{3.10a}
$$

$$
= (1-q^\varepsilon)\left[\begin{array}{c} q^2, 1/bc, 1/bd, 1/cd \\ q/b, q/c, q/d, q^{-1}/bcd \end{array}; q\right]_\infty \tag{3.10b}
$$

where the $\varepsilon = -1$ case corresponds to the Bailey formula (3.5b).

Both formulae are transformed under $\mathcal{R} = \mathcal{S}$ and $\mathcal{S}^{-1}$, respectively, to:

$$
\Theta\left[\begin{array}{ccc} 2, & 2, & 2 \\ b, & c, & d \end{array}; \varepsilon\right] \underset{\varepsilon=2,3,4}{=\!=\!=\!=}(1-q^{3-\varepsilon})\left[\begin{array}{c} q^2, q^2/bc, q^2/bd, q^2/cd \\ q^2/b, q^2/c, q^2/d, q^2/bcd \end{array}; q\right]_\infty \tag{3.11a}
$$

$$
\Theta\left[\begin{array}{ccc} -2, & -2, & -2 \\ b, & c, & d \end{array}; \varepsilon\right] \underset{\varepsilon=-2,-3,-4}{=\!=\!=\!=} q^{7+2\varepsilon}(1-q^{-3-\varepsilon})\left[\begin{array}{c} b, bq, c, cq, d, dq \\ bcdq^3, bcdq^4 \end{array}\right] \tag{3.11b}
$$

$$
\times \left[\begin{array}{c} q^2, q^{-2}/bc, q^{-2}/bd, q^{-2}/cd \\ q^{-2}/b, q^{-2}/c, q^{-2}/d, q^{-2}/bcd \end{array}; q\right]_\infty \tag{3.11c}
$$

The evaluation from the above example

$$
\Theta\left[\begin{array}{ccc} 0, & 0, & 0 \\ b, & c, & d \end{array}; 0\right] = 0 \tag{3.12a}
$$

is not an isolated fact. In general, let

$$
\theta_n(\varepsilon) = {}_n\psi_n\left[\begin{array}{cccc} x_1, & x_2, & \cdots, & x_n \\ 1/x_1, & 1/x_2, & \cdots, & 1/x_n \end{array}; \frac{q^\varepsilon}{x_1 x_2 \cdots x_n}\right]. \tag{3.12b}
$$

Reversing the summation index, we find

$$\theta_n(\varepsilon) \;=\; (-1)^n \, q^\varepsilon \theta_n(-\varepsilon). \tag{3.12c}$$

It yields that

$$\theta_{2n}(\varepsilon) \;=\; q^\varepsilon \theta_{2n}(-\varepsilon) \tag{3.12d}$$

$$\theta_{1+2n}(\varepsilon) \;=\; -\,q^\varepsilon \theta_{1+2n}(-\varepsilon). \tag{3.12e}$$

In particular, we have

$$\theta_{1+2n}(0) \;=\; 0 \tag{3.12f}$$

which includes (3.12a) as special case.

Example 3.2 *Rewriting the transformations from* Example 2.1

$$\Omega_0 \begin{bmatrix} 1, & 1, & 1, & 1; & \varepsilon \\ bq^{-n}, cq^{-n}, dq^{-n}, eq^{-n}; & n \end{bmatrix} \overset{\varepsilon=2,3}{=\!=\!=} \{bcq^{\varepsilon-3+n}\}^n \begin{bmatrix} q^{-2n}, & q/bc \\ b, & c \end{bmatrix} ; q \Big]_n$$

$$\times \;\; {}_4\phi_3 \begin{bmatrix} q^{-n}, & bq^{-n}, & cq^{-n}, & q/de \\ & q^{1-n}/d, q^{1-n}/e, q^{-n}bc \end{bmatrix} ; q \Big]$$

with the above displayed ${}_4\phi_3$ *being replaced by its reversal*

$${}_4\phi_3 \begin{bmatrix} q^{-n}, & bq^{-n}, & cq^{-n}, & q/de \\ & q^{1-n}/d, q^{1-n}/e, q^{-n}bc \end{bmatrix} ; q \Big] = (de/q)^n \begin{bmatrix} q/b, & q/c, & q/de \\ d, & e, & q/bc \end{bmatrix} ; q \Big]_n$$

$$\times \;\; {}_4\phi_3 \begin{bmatrix} q^{-n}, & d, & e, & q/bc \\ & q/b, q/c, q^{-n}de \end{bmatrix} ; q \Big]$$

we can specify (3.6) *as two terminating series-transforms*

$${}_5\psi_5 \begin{bmatrix} q^{-n}, & b, & c, & d, & e & ; & \dfrac{q^{\varepsilon+n}}{bcde} \\ q^{1+n}, & q/b, & q/c, & q/d, & q/e \end{bmatrix} \tag{3.13a}$$

$$\overset{\varepsilon=2,3}{=\!=\!=} \begin{bmatrix} q, \; q/de \\ q/d, \; q/e \end{bmatrix} ; q \Big]_n \; {}_4\phi_3 \begin{bmatrix} q^{-n}, & d, & e, & q/bc \\ & q/b, & q/c, & q^{-n}de \end{bmatrix} ; q \Big] \tag{3.13b}$$

whose limiting versions read as

$${}_4\psi_5 \begin{bmatrix} b, & c, & d, & e & ; & \dfrac{q^\varepsilon}{bcde} \\ 0, & q/b, & q/c, & q/d, & q/e \end{bmatrix} \quad (\varepsilon = 2, \, 3) \tag{3.14a}$$

$$= \begin{bmatrix} q, \; q/de \\ q/d, \; q/e \end{bmatrix} ; q \Big]_\infty \; {}_3\phi_2 \begin{bmatrix} q/bc, & d, & e & ; & \dfrac{q}{de} \\ & q/b, & q/c \end{bmatrix} \tag{3.14b}$$

$$= \begin{bmatrix} q, \; q/bc \\ q/b, \; q/c \end{bmatrix} ; q \Big]_\infty \; {}_3\phi_2 \begin{bmatrix} q/de, & b, & c & ; & \dfrac{q}{bc} \\ & q/d, & q/e \end{bmatrix} \tag{3.14c}$$

*where the last line follows from the symmetry on parameters b, c, d and e, which
is also a direct consequence of Hall's formula* (3.4b).

When $e = 0$, the $_3\phi_2$ reduces to $_2\phi_1$ which may be evaluated by the q-Gauss theorem. In this case, we have two bilateral basic hypergeometric identities

$$_3\psi_3 \left[\begin{array}{ccc} b, & c, & d \\ q/b, & q/c, & q/d \end{array} ; \frac{q^\varepsilon}{bcd} \right] \overset{\varepsilon=1,2}{=\!=\!=} \left[\begin{array}{c} q, \ q/bc, \ q/bd, \ q/cd \\ q/b, \ q/c, \ q/d, \ q/bcd \end{array} ; q \right]_\infty \quad (3.15a)$$

where the $\varepsilon = 1$ case corresponds to the Bailey formula (3.5a).

Both formulae are equivalent each other, under $\mathcal{R}$-operator and transformed under $\mathcal{S}^{-1}$, to:

$$\Theta \left[\begin{array}{ccc} -1, & -1, & -1 \\ b, & c, & d \end{array} ; \varepsilon \right] \overset{\varepsilon=-1,-2}{=\!=\!=\!=} - q^{2+\varepsilon} \left[\begin{array}{c} b, \ c, \ d \\ bcdq^2 \end{array} \right] \quad (3.15b)$$

$$\times \left[\begin{array}{c} q, \ q^{-1}/bc, \ q^{-1}/bd, \ q^{-1}/cd \\ q^{-1}/b, \ q^{-1}/c, \ q^{-1}/d, \ q^{-1}/bcd \end{array} ; q \right]_\infty \quad (3.15c)$$

Two general identities with extra parameters u, v and w

$$_5\psi_5 \left[\begin{array}{ccccc} qu, & qv, & b, & c, & d \\ u, & v, & 1/b, 1/c, 1/d \end{array} ; \frac{q^{-1}}{bcd} \right] = \frac{-q^{-1}(1-quv)}{(1-u)(1-v)} \quad (3.16a)$$

$$\times \left[\begin{array}{c} q, \ 1/bc, \ 1/bd, \ 1/cd \\ q/b, \ q/c, \ q/d, \ q^{-1}/bcd \end{array} ; q \right]_\infty \quad (3.16b)$$

$$_4\psi_4 \left[\begin{array}{cccc} qw, & b, & c, & d \\ w, & q/b, & q/c, & q/d \end{array} ; \frac{q}{bcd} \right] = \left[\begin{array}{c} q, \ q/bc, \ q/bd, \ q/cd \\ q/b, \ q/c, \ q/d, \ q/bcd \end{array} ; q \right]_\infty \quad (3.16c)$$

may be derived respectively, from the combination of the formulae displayed in (3.10a-3.10b) and (3.15a).

Remark For every integer n, the last two examples allow us to evaluate the bilateral well-poised series

$$\Theta \left[\begin{array}{ccc} n, & n, & n \\ b, & c, & d \end{array} ; \varepsilon + 3n/2 \right], \left\{ \begin{array}{ll} \varepsilon = 0, \pm 1, & n \equiv 0 \pmod 2 \\ \varepsilon = \pm 1/2, & n \equiv 1 \pmod 2 \end{array} \right.$$

in view of the shifted operator $\mathcal{S}$.

Example 3.3 *Rewriting the transformations from Example 2.6*

$$\Omega_0 \left[\begin{array}{cc} 0, \ \ 0, \ \ 1, \ \ 1; \ \ \varepsilon \\ bq^{-n}, cq^{-n}, dq^{-n}, eq^{-n}; \ n \end{array} \right] \overset{\varepsilon=1,2}{=\!=\!=} \{bcq^{n+1}\}^n \left[\begin{array}{cc} q^{-2n}, & 1/bc \\ bq, & cq \end{array} ; q \right]_n$$

$$\times \ _4\phi_3 \left[\begin{array}{cccc} q^{-n}, & bq^{-n}, & cq^{-n}, & q/de \\ & q^{1-n}/d, q^{1-n}/e, q^{1-n}bc \end{array} ; q^\varepsilon \right]$$

with the above displayed $_4\phi_3$ being replaced by its reversal

$$_4\phi_3 \left[\begin{array}{cccc} q^{-n}, & bq^{-n}, & cq^{-n}, & q/de \\ & q^{1-n}/d, q^{1-n}/e, q^{1-n}bc \end{array} ; q^\varepsilon \right] = (deq^{\varepsilon-3})^n \left[\begin{array}{ccc} q/b, & q/c, & q/de \\ d, & e, & 1/bc \end{array} ; q \right]_n$$

$$\times \ _4\phi_3 \left[\begin{array}{cccc} q^{-n}, & d, & e, & 1/bc \\ & q/b, q/c, q^{-n}de & \end{array} ; q^{3-\varepsilon} \right]$$

we can specify (3.6) as two terminating series-transforms

$$_5\psi_5 \left[\begin{array}{ccccc} q^{-n}, & b, & c, & d, & e \\ q^{1+n}, & 1/b, & 1/c, & q/d, & q/e \end{array} ; \frac{q^{\varepsilon+n}}{bcde} \right] \qquad (\varepsilon = 1, \, 2) \qquad (3.17a)$$

$$= \left[\begin{array}{c} q, \, q/de \\ q/d, \, q/e \end{array} ; q \right]_n {}_4\phi_3 \left[\begin{array}{cccc} q^{-n}, & d, & e, & 1/bc \\ & q/b, & q/c, & q^{-n}de \end{array} ; q^{3-\varepsilon} \right] \qquad (3.17b)$$

whose limiting versions read as:

$$_4\psi_5 \left[\begin{array}{ccccc} & b, & c, & d, & e \\ 0, & 1/b, & 1/c, & q/d, & q/e \end{array} ; \frac{q^{\varepsilon}}{bcde} \right] \qquad (\varepsilon = 1, \, 2) \qquad (3.18a)$$

$$= \left[\begin{array}{c} q, \, q/de \\ q/d, \, q/e \end{array} ; q \right]_\infty {}_3\phi_2 \left[\begin{array}{ccc} 1/bc, & d, & e \\ & q/b, & q/c \end{array} ; \frac{q^{3-\varepsilon}}{de} \right]. \qquad (3.18b)$$

Interchanging c and e in (3.18a-3.18b), we have a transformation

$$_4\psi_5 \left[\begin{array}{ccccc} & b, & c, & d, & e \\ 0, & 1/b, & q/c, & q/d, & 1/e \end{array} ; \frac{q^{\varepsilon}}{bcde} \right] \qquad (\varepsilon = 1, \, 2) \qquad (3.19a)$$

$$= \left[\begin{array}{c} q, \, q/cd \\ q/c, \, q/d \end{array} ; q \right]_\infty {}_3\phi_2 \left[\begin{array}{ccc} 1/be, & c, & d \\ & q/b, & q/e \end{array} ; \frac{q^{3-\varepsilon}}{cd} \right] \qquad (3.19b)$$

which can be specified, by setting e = 0, as

$$_3\psi_3 \left[\begin{array}{ccc} b, & c, & d \\ 1/b, & q/c, & q/d \end{array} ; \frac{q^{\varepsilon}}{bcd} \right] \qquad (\varepsilon = 1, \, 2) \qquad (3.20a)$$

$$= \left[\begin{array}{c} q, \, q/cd \\ q/c, \, q/d \end{array} ; q \right]_\infty {}_2\phi_1 \left[\begin{array}{cc} c, \, d \\ q/b \end{array} ; \frac{q^{2-\varepsilon}}{bcd} \right]. \qquad (3.20b)$$

For $\varepsilon = 1$, the $_2\phi_1$ may be evaluated by the q-Gauss theorem. In this case, a bilateral basic hypergeometric identity due to Jain & Verma [15] *is recovered*

$$_3\psi_3 \left[\begin{array}{ccc} b, & c, & d \\ 1/b, & q/c, & q/d \end{array} ; \frac{q}{bcd} \right] = \left[\begin{array}{c} q, \, q/bc, \, q/bd, \, q/cd \\ q/b, \, q/c, \, q/d, \, q/bcd \end{array} ; q \right]_\infty \qquad (3.21a)$$

which is a special case of (3.16c).

For $\varepsilon = 2$, the corresponding bilateral identity reads as

$$_3\psi_3 \left[\begin{array}{ccc} b, & c, & d \\ 1/b, & q/c, & q/d \end{array} ; \frac{q^2}{bcd} \right] = \left\{ 1 - \frac{(1-c)(1-d)}{1-bcd} \right\} \qquad (3.21b)$$

$$\times \left[\begin{array}{c} q, \, q/bc, \, q/bd, \, q/cd \\ q/b, \, q/c, \, q/d, \, q/bcd \end{array} ; q \right]_\infty \qquad (3.21c)$$

which follows from the manipulation on $_2\phi_1$ from (3.20b) by the q-Gauss theorem

$$_2\phi_1 \left[\begin{array}{c} c, \, d \\ q/b \end{array} ; \frac{1}{bcd} \right] = {}_2\phi_1 \left[\begin{array}{c} c, \, d \\ q/b \end{array} ; \frac{q}{bcd} \right]$$

$$+ \frac{(1-c)(1-d)}{bcd(1-q/b)} {}_2\phi_1 \left[\begin{array}{c} cq, \, dq \\ q^2/b \end{array} ; \frac{1}{bcd} \right]$$

$$= \left\{ 1 - \frac{(1-c)(1-d)}{1-bcd} \right\} \left[\begin{array}{c} q, \, q/bc \\ q/b, \, q/bcd \end{array} ; q \right]_\infty$$

in view of $1 = q^k + (1 - q^k)$.

Even for $\varepsilon = 0$, a bilateral evaluation

$$_3\psi_3 \left[\begin{array}{ccc} b, & c, & d \\ 1/b, & q/c, & q/d \end{array} ; \frac{1}{bcd} \right] = \{1 + b\frac{(1-c)(1-d)}{1-bcd}\} \tag{3.21d}$$

$$\times \left[\begin{array}{cccc} q, & q/bc, & q/bd, & q/cd \\ q/b, & q/c, & q/d, & q/bcd \end{array} ; q \right]_\infty \tag{3.21e}$$

may be established by expressing $1 = (1 - bq^k) + bq^k$ and

$$_3\psi_3 \left[\begin{array}{ccc} b, & c, & d \\ 1/b, & q/c, & q/d \end{array} ; \frac{1}{bcd} \right] = b \; _3\psi_3 \left[\begin{array}{ccc} b, & c, & d \\ 1/b, & q/c, & q/d \end{array} ; \frac{q}{bcd} \right]$$

$$+ (1 - b) \; _3\psi_3 \left[\begin{array}{ccc} bq, & c, & d \\ 1/b, & q/c, & q/d \end{array} ; \frac{1}{bcd} \right]$$

in accordance with (3.15a) and (3.21a).

All the formulae displayed in (3.21) are transformed, under $\mathcal{RS}$, $\mathcal{R}$ and $\mathcal{S}^{-1}$ respectively, to:

$$\Theta \left[\begin{array}{ccc} 0, & -1, & -1 \\ b, & c, & d \end{array} ; 0 \right] = \left[\begin{array}{c} cdq \\ cq, \; dq \end{array} \right] q \{1 + b\frac{(1-cq)(1-dq)}{1-bcdq^2}\} \tag{3.22a}$$

$$\times \left[\begin{array}{cccc} q, & 1/bc, & 1/bd, & 1/cd \\ q/b, & q/c, & q/d, & q^{-1}/bcd \end{array} ; q \right]_\infty \tag{3.22b}$$

$$\Theta \left[\begin{array}{ccc} 0, & -1, & -1 \\ b, & c, & d \end{array} ; -1 \right] = \left[\begin{array}{c} cdq \\ cq, \; dq \end{array} \right] \left[\begin{array}{c} q, 1/bc, 1/bd, 1/cd \\ q/b, q/c, q/d, q^{-1}/bcd \end{array} ; q \right]_\infty \tag{3.22c}$$

$$\Theta \left[\begin{array}{ccc} 0, & -1, & -1 \\ b, & c, & d \end{array} ; -2 \right] = \left[\begin{array}{c} cdq \\ cq, \; dq \end{array} \right] q^{-1} \{1 - \frac{(1-cq)(1-dq)}{1-bcdq^2}\} \tag{3.22d}$$

$$\times \left[\begin{array}{cccc} q, & 1/bc, & 1/bd, & 1/cd \\ q/b, & q/c, & q/d, & q^{-1}/bcd \end{array} ; q \right]_\infty \tag{3.22e}$$

$$\Theta \left[\begin{array}{ccc} 1, & 1, & 2 \\ b, & c, & d \end{array} ; 1 \right] = \{1 - \frac{(1-b)(1-c)}{1-bcd/q}\} \tag{3.23a}$$

$$\times \left[\begin{array}{cccc} q, & q/bc, & q^2/bd, & q^2/cd \\ q/b, & q/c, & q^2/d, & q^2/bcd \end{array} ; q \right]_\infty \tag{3.23b}$$

$$\Theta \left[\begin{array}{ccc} 1, & 1, & 2 \\ b, & c, & d \end{array} ; 2 \right] = \left[\begin{array}{cccc} q, & q/bc, & q^2/bd, & q^2/cd \\ q/b, & q/c, & q^2/d, & q^2/bcd \end{array} ; q \right]_\infty \tag{3.23c}$$

$$\Theta \left[\begin{array}{ccc} 1, & 1, & 2 \\ b, & c, & d \end{array} ; 3 \right] = \{1 + d/q\frac{(1-b)(1-c)}{1-bcd/q}\} \tag{3.23d}$$

$$\times \left[\begin{array}{cccc} q, & q/bc, & q^2/bd, & q^2/cd \\ q/b, & q/c, & q^2/d, & q^2/bcd \end{array} ; q \right]_\infty \tag{3.23e}$$

$$\Theta\begin{bmatrix} -1, -1, -2 \\ b, \quad c, \quad d \end{bmatrix}; -1\end{bmatrix} = -q^2 \begin{bmatrix} b, c, d, dq \\ bcdq^2, dq^2 \end{bmatrix} \{1 - \frac{(1-bq)(1-cq)}{1-bcdq^3}\} \quad (3.24a)$$

$$\times \begin{bmatrix} q, & q^{-1}/bc, & q^{-1}/bd, & q^{-1}/cd \\ q^{-1}/b, & q^{-1}/c, & q^{-1}/d, & q^{-1}/bcd \end{bmatrix}; q\end{bmatrix}_\infty \quad (3.24b)$$

$$\Theta\begin{bmatrix} -1, -1, -2 \\ b, \quad c, \quad d \end{bmatrix}; -2\end{bmatrix} = -q \begin{bmatrix} b, c, d, dq \\ bcdq^2, dq^2 \end{bmatrix} \times \quad (3.24c)$$

$$\times \begin{bmatrix} q, & q^{-1}/bc, & q^{-1}/bd, & q^{-1}/cd \\ q^{-1}/b, & q^{-1}/c, & q^{-1}/d, & q^{-1}/bcd \end{bmatrix}; q\end{bmatrix}_\infty \quad (3.24d)$$

$$\Theta\begin{bmatrix} -1, -1, -2 \\ b, \quad c, \quad d \end{bmatrix}; -3\end{bmatrix} = - \begin{bmatrix} b, c, d, dq \\ bcdq^2, dq^2 \end{bmatrix} \{1 + dq\frac{(1-bq)(1-cq)}{1-bcdq^3}\} \quad (3.24e)$$

$$\times \begin{bmatrix} q, & q^{-1}/bc, & q^{-1}/bd, & q^{-1}/cd \\ q^{-1}/b, & q^{-1}/c, & q^{-1}/d, & q^{-1}/bcd \end{bmatrix}; q\end{bmatrix}_\infty \quad (3.24f)$$

In order to obtain another group of bilateral evaluations, we now make further reformulation on (3.18a-3.18b).

When $\varepsilon = 1$, the $_3\phi_2$ from (3.18b) may be reformulated by (3.4a), as

$$_3\phi_2\begin{bmatrix} 1/bc, & d, & e \\ & q/b, & q/c \end{bmatrix}; \frac{q^2}{de}\end{bmatrix} = \begin{bmatrix} q/bd, & q^2/e \\ q/b, & q^2/de \end{bmatrix}; q\end{bmatrix}_\infty$$

$$\times \; _3\phi_2\begin{bmatrix} q/ce, & bq, & d \\ & q/c, & q^2/e \end{bmatrix}; q/bd\end{bmatrix}$$

which leads (3.18b) to the following transformation

$$_4\psi_5\begin{bmatrix} & b, & c, & d, & e \\ 0, & 1/b, & 1/c, & q/d, & q/e \end{bmatrix}; \frac{q}{bcde}\end{bmatrix} \quad (3.25a)$$

$$= \frac{1-q/de}{1-q/e} \begin{bmatrix} q, & q/bd \\ q/b, & q/d \end{bmatrix}; q\end{bmatrix}_\infty \quad (3.25b)$$

$$\times \quad _3\phi_2\begin{bmatrix} q/ce, & bq, & d \\ & q/c, & q^2/e \end{bmatrix}; q/bd\end{bmatrix}. \quad (3.25c)$$

While $\varepsilon = 2$, the corresponding transformation reads as

$$_4\psi_5\begin{bmatrix} & b, & c, & d, & e \\ 0, & 1/b, & 1/c, & q/d, & q/e \end{bmatrix}; \frac{q^2}{bcde}\end{bmatrix} \quad (3.25d)$$

$$= \frac{1-q(d+e-de)/de}{(1-q/d)(1-q/e)} \begin{bmatrix} q, & 1/bc \\ q/b, & q/c \end{bmatrix}; q\end{bmatrix}_\infty \quad (3.25e)$$

$$\times \quad _4\phi_3\begin{bmatrix} q/de, bq, cq, q^2(d+e-de)/de \\ q^2/d, q^2/e, \quad q(d+e-de)/de \end{bmatrix}; 1/bc\end{bmatrix} \quad (3.25f)$$

which follows from splitting (3.18b) *by*

$$1 = \frac{d}{d-e}(1 - eq^k) - \frac{e}{d-e}(1 - dq^k)$$

and transforming the $_3\phi_2$*'s by* (3.4b) *as follows:*

$$
_3\phi_2\left[\begin{array}{ccc} 1/bc, & d, & e \\ & q/b, & q/c \end{array}; \frac{q}{de}\right]
$$

$$
= \frac{d(1-e)}{d-e}\,_3\phi_2\left[\begin{array}{ccc} 1/bc, & d, & eq \\ & q/b, & q/c \end{array}; \frac{q}{de}\right]
$$

$$
- \frac{e(1-d)}{d-e}\,_3\phi_2\left[\begin{array}{ccc} 1/bc, & dq, & e \\ & q/b, & q/c \end{array}; \frac{q}{de}\right]
$$

$$
= \frac{d(1-e)}{d-e}\left[\begin{array}{c} q^2/d,\ q/e,\ 1/bc \\ q/b,\ q/c,\ q/de \end{array}; q\right]_\infty \,_3\phi_2\left[\begin{array}{ccc} q/de, & bq, & cq \\ & q^2/d, & q/e \end{array}; 1/bc\right]
$$

$$
- \frac{e(1-d)}{d-e}\left[\begin{array}{c} q/d,\ q^2/e,\ 1/bc \\ q/b,\ q/c,\ q/de \end{array}; q\right]_\infty \,_3\phi_2\left[\begin{array}{ccc} q/de, & bq, & cq \\ & q/d, & q^2/e \end{array}; 1/bc\right]
$$

$$
= \left\{1 - q\frac{d+e-de}{de}\right\}\left[\begin{array}{ccc} q^2/d, & q^2/e, & 1/bc \\ q/b, & q/c, & q/de \end{array}; q\right]_\infty
$$

$$
\times\ _4\phi_3\left[\begin{array}{cccc} q/de, & bq, & cq, & q^2(d+e-de)/de \\ q^2/d, & q^2/e, & q(d+e-de)/de \end{array}; 1/bc\right].
$$

For $e = 0$, *the* $_3\phi_2$ *from* (3.25c) *and the* $_4\phi_3$ *from* (3.25f) *reduce, respectively, to* $_2\phi_1$ *which may be evaluated by the q-Gauss theorem. In this case, we have two bilateral basic hypergeometric identities:*

$$
_3\psi_3\left[\begin{array}{ccc} b, & c, & d \\ 1/b, & 1/c, & q/d \end{array}; \frac{q^\varepsilon}{bcd}\right] \overset{\varepsilon=0,1}{=\!=\!=} d^{\varepsilon-1}\left[\begin{array}{c} q,\ 1/bc,\ q/bd,\ q/cd \\ q/b,\ q/c,\ q/d,\ 1/bcd \end{array}; q\right]_\infty \quad (3.26)
$$

Both formulae are transformed under $\mathcal{RS}$, $\mathcal{R}$ *and* $\mathcal{S}^{-1}$ *respectively, to:*

$$
\Theta\left[\begin{array}{ccc} 0, & 0, -1 \\ b, & c, & d \end{array}; \varepsilon\right] \overset{\varepsilon=0,-1}{=\!=\!=} \frac{d^{-\varepsilon}}{1-dq}\left[\begin{array}{c} q,\ 1/bc,\ 1/bd,\ 1/cd \\ q/b,\ q/c,\ q/d,\ q^{-1}/bcd \end{array}; q\right]_\infty \quad (3.26a)
$$

$$
\Theta\left[\begin{array}{ccc} 1, & 2, & 2 \\ b, & c, & d \end{array}; \varepsilon\right] \overset{\varepsilon=2,3}{=\!=\!=} b^{2-\varepsilon}\left[\begin{array}{c} q,\ q^2/bc,\ q^2/bd,\ q^2/cd \\ q/b,\ q^2/c,\ q^2/d,\ q^2/bcd \end{array}; q\right]_\infty \quad (3.26b)
$$

$$
\Theta\left[\begin{array}{ccc} -1, -2, -2 \\ b, & c, & d \end{array}; \varepsilon\right] \overset{\varepsilon=-2,-3}{=\!=\!=} -q^{5+2\varepsilon}b^{3+\varepsilon}\left[\begin{array}{ccc} & cdq^2 \\ bq, & cq^2, & dq^2 \end{array}\right] \quad (3.26c)
$$

$$
\times\ \left[\begin{array}{c} q,\ q^{-1}/bc,\ q^{-1}/bd,\ q^{-1}/cd \\ q/b,\ q/c,\ q/d,\ q^{-3}/bcd \end{array}; q\right]_\infty \quad (3.26d)
$$

Remark For every integer n, this example allows us to evaluate the bilateral almost-poised series

$$\Theta \begin{bmatrix} n, & n, & n+1 \\ b, & c, & d \end{bmatrix} ; \varepsilon + 3n/2 \end{bmatrix}, \begin{cases} \varepsilon = 0,\, 1, & n \equiv 0 \pmod 2 \\ \varepsilon = \pm 1/2,\, 3/2, & n \equiv 1 \pmod 2 \end{cases}$$

$$\Theta \begin{bmatrix} n, & n, & n-1 \\ b, & c, & d \end{bmatrix} ; \varepsilon + 3n/2 \end{bmatrix}, \begin{cases} \varepsilon = 0,\, -1, & n \equiv 0 \pmod 2 \\ \varepsilon = \pm 1/2,\, -3/2, & n \equiv 1 \pmod 2 \end{cases}$$

in view of the shifted operator $\mathcal{S}$.

Example 3.4 *Rewriting the transformations from* Example 2.8

$$\Omega_0 \begin{bmatrix} 0, & -1, & 1, & 1; & 1 \\ bq^{-n}, & cq^{-n}, & dq^{-n}, & eq^{-n}; & n \end{bmatrix}$$

$$= \{bcq^{n+1}\}^n \begin{bmatrix} q^{-2n}, & 1/bc \\ bq, & cq \end{bmatrix} ; q \end{bmatrix}_n$$

$$\times\ {}_5\phi_4 \begin{bmatrix} q^{-n}, & bq^{-n}, & cq^{-n}, & q/de, & q^{-n}/c \\ & q^{1-n}/d, & q^{1-n}/e, & q^{1-n}bc, & q^{-1-n}/c \end{bmatrix} ; q \end{bmatrix}$$

with the above displayed ${}_5\phi_4$ being replaced by its reversal

$${}_5\phi_4 \begin{bmatrix} q^{-n}, & bq^{-n}, & cq^{-n}, & q/de, & q^{-n}/c \\ & q^{1-n}/d, & q^{1-n}/e, & q^{1-n}bc, & q^{-1-n}/c \end{bmatrix} ; q \end{bmatrix}$$

$$= (de/q)^n \begin{bmatrix} qc, & q/b, & q/c, & q/de \\ q^2c, & d, & e, & 1/bc \end{bmatrix} ; q \end{bmatrix}_n$$

$$\times\ {}_5\phi_4 \begin{bmatrix} q^{-n}, & q^2c, & d, & e, & 1/bc \\ & qc, & q/b, & q/c, & q^{-n}de \end{bmatrix} ; q \end{bmatrix}$$

we can specify (3.6) as a terminating series-transform

$${}_5\psi_5 \begin{bmatrix} q^{-n}, & b, & c, & d, & e & ; & \dfrac{q^{1+n}}{bcde} \\ q^{1+n}, & 1/b, & 1/qc, & q/d, & q/e & & \end{bmatrix} \tag{3.27a}$$

$$= \begin{bmatrix} q, q/de \\ q/d, q/e \end{bmatrix} ; q \end{bmatrix}_n {}_5\phi_4 \begin{bmatrix} q^{-n}, q^2c, & d, & e, & 1/bc \\ & qc, & q/b, & q/c, & q^{-n}de \end{bmatrix} ; q \end{bmatrix} \tag{3.27b}$$

whose limiting version reads as

$${}_4\psi_5 \begin{bmatrix} & b, & c, & d, & e & ; & \dfrac{q}{bcde} \\ 0, & 1/b, & 1/qc, & q/d, & q/e & & \end{bmatrix} \tag{3.28a}$$

$$= \begin{bmatrix} q, q/de \\ q/d, q/e \end{bmatrix} ; q \end{bmatrix}_\infty {}_4\phi_3 \begin{bmatrix} 1/bc, & d, & e, & q^2c \\ & q/b, & q/c, & qc \end{bmatrix} ; \dfrac{q}{de} \end{bmatrix}. \tag{3.28b}$$

When $b = 0$, this transformation may be restated as

$${}_3\psi_3 \begin{bmatrix} b, & c, & d & ; & \dfrac{q}{bcd} \\ q/b, & q/c, & 1/qd & & \end{bmatrix}$$

$$= \begin{bmatrix} q, q/bc \\ q/b, q/c \end{bmatrix} ; q \end{bmatrix}_\infty {}_3\phi_2 \begin{bmatrix} b, c, q^2d \\ q/d, qd \end{bmatrix} ; \dfrac{1}{bcd} \end{bmatrix}.$$

From

$$1 - dq^{k+1} = \frac{qd - c}{b - c}(1 - bq^k) + \frac{b - dq}{b - c}(1 - cq^k),$$

evaluating

$$
{}_3\phi_2 \left[\begin{array}{cc} b, & c, & q^2d \\ & q/d, & qd \end{array} ; \frac{1}{bcd} \right] = \frac{(qd - c)(1 - b)}{(1 - qd)(b - c)} \; {}_2\phi_1 \left[\begin{array}{c} qb, & c \\ q/d \end{array} ; 1/bcd \right]
$$

$$
+ \frac{(b - dq)(1 - c)}{(1 - qd)(b - c)} \; {}_2\phi_1 \left[\begin{array}{c} b, & cq \\ q/d \end{array} ; 1/bcd \right]
$$

$$
= \{1 - \frac{(1 - b)(1 - c)}{(1 - bcd)(1 - dq)}\} \left[\begin{array}{c} q/bd, & q/cd \\ q/d, & q/bcd \end{array} ; q \right]_\infty
$$

by the q-Gauss theorem, we get a bilateral basic hypergeometric identity:

$$
{}_3\psi_3 \left[\begin{array}{ccc} b, & c, & d \\ q/b, & q/c, & 1/qd \end{array} ; \frac{q}{bcd} \right] = \{1 - \frac{(1 - b)(1 - c)}{(1 - bcd)(1 - dq)}\} \qquad (3.29a)
$$

$$
\times \left[\begin{array}{cccc} q, & q/bc, & q/bd, & q/cd \\ q/b, & q/c, & q/d, & q/bcd \end{array} ; q \right]_\infty \qquad (3.29b)
$$

Another bilateral evaluation

$$
{}_3\psi_3 \left[\begin{array}{ccc} b, & c, & d \\ q/b, & q/c, & 1/qd \end{array} ; \frac{1}{bcd} \right] = \{1 - qd^2 \frac{(1 - b)(1 - c)}{(1 - bcd)(1 - dq)}\} \qquad (3.29c)
$$

$$
\times \left[\begin{array}{cccc} q, & q/bc, & q/bd, & q/cd \\ q/b, & q/c, & q/d, & q/bcd \end{array} ; q \right]_\infty \qquad (3.29d)
$$

may be established by expressing $1 = (1 - dq^k) + dq^k$ *and*

$$
{}_3\psi_3 \left[\begin{array}{ccc} b, & c, & d \\ q/b, & q/c, & 1/qd \end{array} ; \frac{1}{bcd} \right] = d \; {}_3\psi_3 \left[\begin{array}{ccc} b, & c, & d \\ q/b, & q/c, & 1/qd \end{array} ; \frac{q}{bcd} \right]
$$

$$
+ (1 - d) \; {}_3\psi_3 \left[\begin{array}{ccc} b, & c, & qd \\ q/b, & q/c, & 1/qd \end{array} ; \frac{1}{bcd} \right]
$$

in accordance with (3.21d-3.21e) and (3.29a-3.29b).

Both formulae stated in (3.29) are transformed under $\mathcal{RS}$, $\mathcal{R}$, $\mathcal{S}^{-1}$, $\mathcal{S}$, and $\mathcal{RS}^2$ respectively, to:

$$
\Theta \left[\begin{array}{ccc} 1, & -1, & -1 \\ b, & c, & d \end{array} ; \varepsilon \right] \xLongequal{\varepsilon = 0, -1} b^{2\varepsilon} \{1 - (q/b^2)^{1+\varepsilon} \frac{(1 - b)(1 - bcdq)}{(1 - cq)(1 - dq)}\} \qquad (3.30a)
$$

$$
\times \left[\begin{array}{cccc} q, & q/bc, & q/bd, & q^{-1}/cd \\ q/b, & q/c, & q/d, & q^{-1}/bcd \end{array} ; q \right]_\infty \qquad (3.30b)
$$

$$
\Theta \left[\begin{array}{ccc} 1, & 1, & 3 \\ b, & c, & d \end{array} ; \varepsilon \right] \xLongequal{\varepsilon = 2, 3} \{1 - (d^2/q^3)^{\varepsilon - 2} \frac{(1 - b)(1 - c)}{(1 - bcd/q^2)(1 - d/q)}\} \qquad (3.30c)
$$

$$
\times \left[\begin{array}{cccc} q, & q/bc, & q^3/bd, & q^3/cd \\ q/b, & q/c, & q^3/d, & q^3/bcd \end{array} ; q \right]_\infty \qquad (3.30d)
$$

$$\Theta\begin{bmatrix} -1, -1, -3 \\ b, \quad c, \quad d \end{bmatrix}; \varepsilon \Big] \xlongequal{\varepsilon = -2, -3} \{1 - (d^2q^3)^{2+\varepsilon} \frac{(1 - bcdq^3)(1 - dq^2)}{(1 - bq)(1 - cq)}\} \quad (3.30e)$$

$$\times \frac{q^{-2}(d^2q^4)^{-2-\varepsilon}}{[dq^2, dq^3]} \begin{bmatrix} q, \ q^{-1}/bc, \ q^{-1}/bd, \ q^{-1}/cd \\ q/b, \ q/c, \ q/d, \ q^{-3}/bcd \end{bmatrix}; q \Big]_\infty \quad (3.30f)$$

$$\Theta\begin{bmatrix} 1, \quad 3, \quad 3 \\ b, \quad c, \quad d \end{bmatrix}; \varepsilon \Big] \xlongequal{\varepsilon = 3, 4} b^{6-2\varepsilon}\{1 - (q/b^2)^{4-\varepsilon} \frac{(1 - b)(1 - bcd/q^3)}{(1 - c/q)(1 - d/q)}\} \quad (3.30g)$$

$$\times \begin{bmatrix} q, \ q^3/bc, \ q^3/bd, \ q^3/cd \\ q/b, \ q^3/c, \ q^3/d, \ q^3/bcd \end{bmatrix}; q \Big]_\infty \quad (3.30h)$$

$$\Theta\begin{bmatrix} -1, -3, -3 \\ b, \quad c, \quad d \end{bmatrix}; \varepsilon \Big] \xlongequal{\varepsilon = -3, -4} \{1 - (qb^2)^{-4-\varepsilon} \frac{(1 - bq)(1 - bcdq^4)}{(1 - cq^2)(1 - dq^2)}\} \quad (3.30i)$$

$$\times \frac{q^{13+3\varepsilon}b^{7+2\varepsilon}cd}{[bq, cq^3, dq^3]} \begin{bmatrix} q, \ q^{-1}/bc, \ q^{-1}/bd, \ q^{-3}/cd \\ q/b, \ q/c, \ q/d, \ q^{-4}/bcd \end{bmatrix}; q \Big]_\infty \quad (3.30j)$$

Example 3.5 *We can also establish another bilateral transformation from (3.28a-3.28b).*

Writing the $_4\phi_3$ from (3.28b) in terms of the balanced $_3\phi_2$-series according to

$$1 - cq^{k+1} = \frac{qc - e}{d - e}(1 - dq^k) + \frac{d - cq}{d - e}(1 - eq^k)$$

and then transforming them by (3.4b)

$$_4\phi_3\begin{bmatrix} 1/bc, & d, & e, & q^2c \\ & q/b, & q/c, & qc \end{bmatrix}; \frac{q}{de} \Big]$$

$$= \frac{(qc - e)(1 - d)}{(1 - qc)(d - e)} \, _3\phi_2\begin{bmatrix} 1/bc, & qd, & e \\ & q/b, & q/c \end{bmatrix}; \frac{q}{de} \Big]$$

$$+ \frac{(d - cq)(1 - e)}{(1 - qc)(d - e)} \, _3\phi_2\begin{bmatrix} 1/bc, & d, & qe \\ & q/b, & q/c \end{bmatrix}; \frac{q}{de} \Big]$$

$$= \frac{(qc - e)(1 - d)}{(1 - qc)(d - e)} \begin{bmatrix} 1/bc, \ q/d, \ q^2/e \\ q/b, \ q/c, \ q/de \end{bmatrix}; q \Big]_\infty \, _3\phi_2\begin{bmatrix} q/de, & qb, & qc \\ & q/d, & q^2/e \end{bmatrix}; 1/bc \Big]$$

$$+ \frac{(d - cq)(1 - e)}{(1 - qc)(d - e)} \begin{bmatrix} 1/bc, \ q^2/d, \ q/e \\ q/b, \ q/c, \ q/de \end{bmatrix}; q \Big]_\infty \, _3\phi_2\begin{bmatrix} q/de, & qb, & qc \\ & q^2/d, & q/e \end{bmatrix}; 1/bc \Big]$$

$$= \{1 - q\frac{d + e - de - cq}{(1 - qc)de}\} \begin{bmatrix} 1/bc, & q^2/d, & q^2/e \\ q/b, & q/c, & q/de \end{bmatrix}; q \Big]_\infty$$

$$\times \, _4\phi_3\begin{bmatrix} q/de, \ qb, \ qc, q^2(d + e - de - cq)/(1 - qc)de \\ q^2/d, \ q^2/e, \ q(d + e - de - cq)/(1 - qc)de \end{bmatrix}; 1/bc \Big]$$

we can reformulate (3.28a-3.28b) from Example 3.4 *as another transformation*

$$
{}_4\psi_5\left[\begin{array}{ccccc} b, & c, & d, & e \\ 0, & 1/b, & 1/qc, & q/d, & q/e \end{array}; \frac{q}{bcde}\right] \tag{3.31a}
$$

$$
= \frac{1-q(d+e-de-cq)/(1-qc)de}{(1-q/d)(1-q/e)}\left[\begin{array}{cc} q, & 1/bc \\ q/b, & q/c \end{array}; q\right]_\infty \tag{3.31b}
$$

$$
\times \ {}_4\phi_3\left[\begin{array}{cccc} q/de, & qb, & qc, & q^2(d+e-de-cq)/(1-qc)de \\ q^2/d, & q^2/e, & q(d+e-de-cq)/(1-qc)de \end{array}; 1/bc\right]. \tag{3.31c}
$$

When $e = 0$, *it reduces to a bilateral basic hypergeometric identity in view of the q-Gauss theorem*

$$
{}_3\psi_3\left[\begin{array}{ccc} b, & c, & d \\ q/b, & 1/c, & 1/qd \end{array}; \frac{1}{bcd}\right] \tag{3.32a}
$$

$$
= \frac{1-qd/b}{1-qd}\left[\begin{array}{c} q, q/bc, q/bd, 1/cd \\ q/b, q/c, q/d, 1/bcd \end{array}; q\right]_\infty. \tag{3.32b}
$$

It is transformed, under $\mathcal{R} = \mathcal{S}$, *to:*

$$
\Theta\left[\begin{array}{ccc} 1, & 2, & 3 \\ b, & c, & d \end{array}; 3\right] = \frac{1-d/bq}{1-d/q}\left[\begin{array}{c} q, q^2/bc, q^3/bd, q^3/cd \\ q/b, q^2/c, q^3/d, q^3/bcd \end{array}; q\right]_\infty \tag{3.32c}
$$

Example 3.6 *Rewriting the transformations from* Example 2.27

$$
\Omega_0\left[\begin{array}{ccccc} 0, & 0, & 1, & 2; & 2 \\ bq^{-n}, & cq^{-n}, & dq^{-n}, & eq^{-n}; & n \end{array}\right]
$$

$$
= \ \{q^{n+1}bc\}^n \left[\begin{array}{cc} q^{-2n}, & 1/bc \\ bq, & cq \end{array}; q\right]_n
$$

$$
\times \ {}_5\phi_4\left[\begin{array}{ccccc} q^{-n}, & bq^{-n}, & cq^{-n}, & q^2/de, & q^{-n}e \\ & q^{1-n}/d, & q^{2-n}/e, & q^{1-n}bc, & q^{-1-n}e \end{array}; q\right]
$$

with the above displayed ${}_5\phi_4$ *being replaced by its reversal*

$$
{}_5\phi_4\left[\begin{array}{ccccc} q^{-n}, & bq^{-n}, & cq^{-n}, & q^2/de, & q^{-n}e \\ & q^{1-n}/d, & q^{2-n}/e, & q^{1-n}bc, & q^{-1-n}e \end{array}; q\right]
$$

$$
= \ (q^{-2}de)^n \left[\begin{array}{cccc} q/e, & q/b, & q/c, & q^2/de \\ q^2/e, & d, & e/q, & 1/bc \end{array}; q\right]_n
$$

$$
\times \ {}_5\phi_4\left[\begin{array}{ccccc} q^{-n}, & q^2/e, & d, & e/q, & 1/bc \\ & q/e, & q/b, & q/c, & q^{-1-n}de \end{array}; q\right]
$$

we can specify (3.6) as a terminating series-transform

$$
{}_5\psi_5\left[\begin{array}{ccccc} q^{-n}, & b, & c, & d, & e \\ q^{1+n}, & 1/b, & 1/c, & q/d, & q^2/e \end{array}; \frac{q^{2+n}}{bcde}\right] \tag{3.33a}
$$

$$
= \left[\begin{array}{c} q, q^2/de \\ q/d, q^2/e \end{array}; q\right]_n {}_5\phi_4\left[\begin{array}{ccccc} q^{-n}, & q^2/e, & d, & e/q, & 1/bc \\ & q/e, & q/b, & q/c, & q^{-1-n}de \end{array}; q\right] \tag{3.33b}
$$

whose limiting version reads as

$$_4\psi_5\left[\begin{matrix} & b, & c, & d, & e \\ 0, & 1/b, & 1/c, & q/d, & q^2/e \end{matrix} \; ; \; \frac{q^2}{bcde}\right] \qquad (3.34a)$$

$$= \left[\begin{matrix} q, & q^2/de \\ q/d, & q^2/e \end{matrix} \; ; \; q\right]_\infty {}_4\phi_3\left[\begin{matrix} 1/bc, & d, & e/q, & q^2/e \\ & q/b, & q/c, & q/e \end{matrix} \; ; \; \frac{q^2}{de}\right]. \qquad (3.34b)$$

When $b = 0$, this transformation may be restated as

$$_3\psi_3\left[\begin{matrix} b, & c, & d \\ 1/b, & q/c, & q^2/d \end{matrix} \; ; \; \frac{q^2}{bcd}\right]$$

$$= \left[\begin{matrix} q, & q^2/cd \\ q/c, & q^2/d \end{matrix} \; ; \; q\right]_\infty {}_3\phi_2\left[\begin{matrix} c, & d/q, & q^2/d \\ & q/b, & q/d \end{matrix} \; ; \; \frac{q}{bcd}\right].$$

From

$$1 - q^{k+1}/d = (1 - q^k/b) + \frac{d - bq}{bd}\, q^k$$

evaluating

$$_3\phi_2\left[\begin{matrix} c, & d/q, & q^2/d \\ & q/b, & q/d \end{matrix} \; ; \; \frac{q}{bcd}\right] \quad = \quad \frac{1 - 1/b}{1 - q/d} \; {}_2\phi_1\left[\begin{matrix} c, & d/q \\ & 1/b \end{matrix} \; ; \; q/bcd\right]$$

$$+ \quad \frac{d - bq}{bd(1 - q/d)} \; {}_2\phi_1\left[\begin{matrix} c, & d/q \\ & q/b \end{matrix} \; ; \; q^2/bcd\right]$$

$$= \quad \{1 + \frac{(1-c)d/q}{1 - bcd/q}\}\left[\begin{matrix} q/bc, & q^2/bd \\ q/b, & q^2/bcd \end{matrix} \; ; \; q\right]_\infty$$

by the q-Gauss theorem, we get a bilateral basic hypergeometric identity:

$$_3\psi_3\left[\begin{matrix} b, & c, & d \\ 1/b, & q/c, & q^2/d \end{matrix} \; ; \; \frac{q^2}{bcd}\right] = \{1 + \frac{(1-c)d/q}{1 - bcd/q}\} \qquad (3.35a)$$

$$\times \left[\begin{matrix} q, & q/bc, & q^2/bd, & q^2/cd \\ q/b, & q/c, & q^2/d, & q^2/bcd \end{matrix} \; ; \; q\right]_\infty \qquad (3.35b)$$

Another bilateral evaluation

$$_3\psi_3\left[\begin{matrix} b, & c, & d \\ 1/b, & q/c, & q^2/d \end{matrix} \; ; \; \frac{q}{bcd}\right] = \{1 + \frac{b(1-c)}{1 - bcd/q}\} \qquad (3.35c)$$

$$\times \left[\begin{matrix} q, & q/bc, & q^2/bd, & q^2/cd \\ q/b, & q/c, & q^2/d, & q^2/bcd \end{matrix} \; ; \; q\right]_\infty \qquad (3.35d)$$

may be established by expressing $1 = (1 - q^{1+k}/d) + q^{1+k}/d$ and

$$_3\psi_3\left[\begin{matrix} b, & c, & d \\ 1/b, & q/c, & q^2/d \end{matrix} \; ; \; \frac{q}{bcd}\right] = q/d \; {}_3\psi_3\left[\begin{matrix} b, & c, & d \\ 1/b, & q/c, & q^2/d \end{matrix} \; ; \; \frac{q^2}{bcd}\right]$$

$$+ (1 - q/d) \; {}_3\psi_3\left[\begin{matrix} b, & c, & d \\ 1/b, & q/c, & q/d \end{matrix} \; ; \; \frac{q}{bcd}\right]$$

in accordance with (3.35a-b) and (3.21a).

Both formulae stated in (3.35) are invariant under $\mathcal{R}$ and transformed under $\mathcal{S}^{-1}$ respectively, to:

$$\Theta\begin{bmatrix} 1, & -1, & -2 \\ b, & c, & d \end{bmatrix};\varepsilon \Bigg]_{\overline{\overline{\varepsilon=-1,-2}}} \{1+(qd/b)^{1+\varepsilon}\frac{1-bcdq^2}{b(1-cq)}\}\frac{(b/d)^{1+\varepsilon}}{1-dq^2} \qquad (3.36a)$$

$$\times \begin{bmatrix} q, & 1/bc, & 1/bd, & q^{-1}/cd \\ q/b, & q/c, & q/d, & q^{-2}/bcd \end{bmatrix};q \Bigg]_{\infty} \qquad (3.36b)$$

Remark For every integer n, Example 3.5 & 3.6 allow us to evaluate the bilateral almost-poised series

$$\Theta\begin{bmatrix} n+1, & n, & n-1 \\ b, & c, & d \end{bmatrix};\varepsilon+3n/2 \Bigg], \qquad \begin{cases} \varepsilon=0, & n\equiv 0 \pmod 2 \\ \varepsilon=\pm 1/2, & n\equiv 1 \pmod 2 \end{cases}$$

in view of the shifted operator $\mathcal{S}$.

Example 3.7 *We can also establish another bilateral transformation from (3.34a-3.34b).*

Writing the $_4\phi_3$ from (3.34b) in terms of the balanced $_3\phi_2$-series according to

$$1-q^{k+1}/e = \frac{1-q^2/e^2}{1-qd/e}(1-dq^k)-qd/e\frac{1-q/de}{1-qd/e}(1-eq^{k-1}),$$

and then transforming them by (3.4b)

$$_4\phi_3\begin{bmatrix} 1/bc, & d, & e/q, & q^2/e \\ & q/b, & q/c, & q/e \end{bmatrix};\frac{q^2}{de}\Bigg]$$

$$= \frac{(1-d)(1-q^2/e^2)}{(1-q/e)(1-qd/e)}\,_3\phi_2\begin{bmatrix} 1/bc, & qd, & e/q \\ & q/b, & q/c \end{bmatrix};\frac{q^2}{de}\Bigg]$$

$$+d\frac{1-q/de}{1-qd/e}\,_3\phi_2\begin{bmatrix} 1/bc, & d, & e \\ & q/b, & q/c \end{bmatrix};\frac{q^2}{de}\Bigg]$$

$$= \frac{(1-d)(1-q^2/e^2)}{(1-q/e)(1-qd/e)}\begin{bmatrix} 1/bc, & q/d, & q^3/e \\ q/b, & q/c, & q^2/de \end{bmatrix};q \Bigg]_{\infty}$$

$$\times\,_3\phi_2\begin{bmatrix} q^2/de, & qb, & qc \\ & q/d, & q^3/e \end{bmatrix};1/bc \Bigg]$$

$$+d\frac{1-q/de}{1-qd/e}\begin{bmatrix} 1/bc, & q^2/d, & q^2/e \\ q/b, & q/c, & q^2/de \end{bmatrix};q \Bigg]_{\infty}$$

$$\times\,_3\phi_2\begin{bmatrix} q^2/de, & qb, & qc \\ & q^2/d, & q^2/e \end{bmatrix};1/bc \Bigg]$$

$$= \{1-q\frac{q+e-de}{de}\}\begin{bmatrix} 1/bc, & q^2/d, & q^3/e \\ q/b, & q/c, & q^2/de \end{bmatrix};q \Bigg]_{\infty}$$

$$\times\,_4\phi_3\begin{bmatrix} q^2/de, & qb, & qc, & q^2(q+e-de)/de \\ q^2/d, & q^3/e, & q(q+e-de)/de \end{bmatrix};1/bc \Bigg]$$

we can reformulate (3.34a-3.34b) from Example 3.6 as another transformation

$$_4\psi_5 \left[\begin{array}{cccc} b, & c, & d, & e \\ 0, & 1/b, & 1/c, & q/d, \quad q^2/e \end{array} ; \frac{q^2}{bcde} \right] \tag{3.37a}$$

$$= \frac{1 - q^{\frac{q+e-de}{de}}}{(1 - q/d)(1 - q^2/e)} \left[\begin{array}{cc} q, & 1/bc \\ q/b, & q/c \end{array} ; q \right]_\infty \tag{3.37b}$$

$$\times \quad _4\phi_3 \left[\begin{array}{c} q^2/de, \; qb, \; qc, \; q^2(q+e-de)/de \\ q^2/d, \; q^3/e, \quad q(q+e-de)/de \end{array} ; 1/bc \right]. \tag{3.37c}$$

When $d = 0$, it reduces a bilateral basic hypergeometric identity in view of the q-Gauss theorem.

$$_3\psi_3 \left[\begin{array}{ccc} b, & c, & d \\ 1/b, & 1/c, & q^2/d \end{array} ; \frac{q}{bcd} \right] = (1 + q/d) \left[\begin{array}{c} q, \; 1/bc, \; q^2/bd, \; q^2/cd \\ q/b, \; q/c, \; q^2/d, \; q/bcd \end{array} ; q \right]_\infty \tag{3.38}$$

It is transformed, under $\mathcal{RS}$, $\mathcal{R}$ and $\mathcal{S}^{-1}$ respectively, to:

$$\Theta \left[\begin{array}{ccc} 0, & 0, & -2 \\ b, & c, & d \end{array} ; -1 \right] = \frac{1 + dq}{1 - dq^2} \left[\begin{array}{c} q, \; 1/bc, \; 1/bd, \; 1/cd \\ q/b, \; q/c, \; q/d, \; q^{-1}/bcd \end{array} ; q \right]_\infty \tag{3.38a}$$

$$\Theta \left[\begin{array}{ccc} 0, & 2, & 2 \\ b, & c, & d \end{array} ; 2 \right] = (1 + 1/b) \left[\begin{array}{c} q, \; q^2/bc, \; q^2/bd, \; q^2/cd \\ q/b, \; q^2/c, \; q^2/d, \; q^2/bcd \end{array} ; q \right]_\infty \tag{3.38b}$$

$$\Theta \left[\begin{array}{ccc} 0, & -2, & -2 \\ b, & c, & d \end{array} ; -2 \right] = \frac{-(1 + 1/b)\, cdq^2}{(1 - cq^2)(1 - dq^2)} \tag{3.38c}$$

$$\times \left[\begin{array}{c} q, \; 1/bc, \; 1/bd, \; q^{-2}/cd \\ q/b, \; q/c, \; q/d, \; q^{-2}/bcd \end{array} ; q \right]_\infty \tag{3.38d}$$

Remark For every integer n, Example 3.4 & 3.7 allow us to evaluate the bilateral almost-poised series

$$\Theta \left[\begin{array}{ccc} n, & n, & n+2 \\ b, & c, & d \end{array} ; \varepsilon + 3n/2 \right], \begin{cases} \varepsilon = 1, & n \equiv 0 \pmod 2 \\ \varepsilon = 1/2, \; 3/2, & n \equiv 1 \pmod 2 \end{cases}$$

$$\Theta \left[\begin{array}{ccc} n, & n, & n-2 \\ b, & c, & d \end{array} ; \varepsilon + 3n/2 \right], \begin{cases} \varepsilon = -1, & n \equiv 0 \pmod 2 \\ \varepsilon = -1/2, \; -3/2, & n \equiv 1 \pmod 2 \end{cases}$$

in view of the shifted operator $\mathcal{S}$.

4 The Rogers-Ramanujan Identities

The Watson Transform (2.1a-2.1b) has been used to give a simple proof of the famous Rogers-Ramanujan identities

(cf. [11, §2.7])

$$\sum_{n=0}^{\infty} \frac{q^{n^2}}{(q;\,q)_n} = \frac{1}{(q;\,q^5)_\infty\,(q^4;\,q^5)_\infty} \tag{4.1a}$$

$$\sum_{n=0}^{\infty} \frac{q^{n(n+1)}}{(q;\,q)_n} = \frac{1}{(q^2;\,q^5)_\infty\,(q^3;\,q^5)_\infty} \tag{4.1b}$$

which may also be derived from (3.3) by letting b, c, d, $e \to \infty$ and then using the Jacobi triple product identity

$$\sum_{n=-\infty}^{+\infty} (-1)^n\,q^{\binom{n}{2}}\,z^n = [q,\,z,\,q/z;\,q]_\infty. \tag{4.2}$$

Following Watson's method, we will try to evaluate the sums of the Rogers-Ramanujan type by combining transformations (1.2) and (2.3).

First, we simplify, gradually, (1.2) by letting $\lambda = 1 - \delta$, $\theta = 1$, and $a = w$, b, c, $\to \infty$. The basic series from (1.2) reduces to

$$_8\phi_7[Eq.1.2] \;\Rightarrow\; {}_1\phi_0\left[\,q^{-\delta-2n+2k}\;;\;q^{\varepsilon+\delta-1+n-k}\,\right] \tag{4.3a}$$

$$= \;(q^{\varepsilon-1-n+k};\,q)_{\delta+2n-2k} \tag{4.3b}$$

which vanishes for $n - k \geq \max\{\varepsilon - 1,\, 2 - \varepsilon - \delta\}$. Substituting this evaluation into (1.2), and then performing replacement $k \to n - k$ on summation index, we have

$$\sum_k[Eq.1.2] \Rightarrow \sum_{k=0}^{n} (-1)^k\,(q^{\varepsilon-1-n+k};\,q)_{\delta+2n-2k}\,\frac{(q^{-\delta-2n};\,q)_{2k}}{(q;\,q)_k}\,q^{k(\varepsilon+\delta-1+n)-\binom{k}{2}} \tag{4.4a}$$

$$= \sum_{k=0}^{\max\{\varepsilon-2,\,1-\varepsilon-\delta\}} (-1)^{n+k}\,\frac{(q^{-\delta-2n};\,q)_{2n-2k}}{(q;\,q)_{n-k}}\,(q^{\varepsilon-1-k};\,q)_{\delta+2k} \tag{4.4b}$$

$$\times\; q^{(n-k)(\varepsilon+\delta-1+n)-\binom{n-k}{2}} \tag{4.4c}$$

$$= \;(-1)^n\,\frac{(q^{1+\delta};\,q)_{2n}}{(q;\,q)_n}\,q^{n(\varepsilon-\delta-2-n)-\binom{n}{2}} \tag{4.4d}$$

$$\times\; \sum_{k=0}^{\max\{\varepsilon-2,\,1-\varepsilon-\delta\}} (q^{-n};\,q)_k\,\frac{(q^{\varepsilon-1-k};\,q)_{\delta+2k}}{(q^{1+\delta};\,q)_{2k}}\,q^{k(2-\varepsilon+\delta+n+k)} \tag{4.4e}$$

79

which leads us to

$$
{}_5\phi_4[Eq.1.2] \quad \Rightarrow \quad \sum_{k=0}^{\delta+2n} \frac{(q^{-\delta-2n}; q)_k}{(q; q)_k} q^{k\,(\varepsilon-1-n+k)} \tag{4.5a}
$$

$$
= \quad (-1)^n \frac{(q^{1+\delta}; q)_{2n}}{(q; q)_n} q^{n\,(\varepsilon-\delta-3)-3\binom{n}{2}} \tag{4.5b}
$$

$$
\times \quad \sum_{k=0}^{\max\{\varepsilon-2,\,1-\varepsilon-\delta\}} (q^{-n}; q)_k \frac{(q^{\varepsilon-1-k}; q)_{\delta+2k}}{(q^{1+\delta}; q)_{2k}} q^{k(2-\varepsilon+\delta+n+k)}. \tag{4.5c}
$$

Second, we reduce (2.3) by letting $\lambda = 1 - \delta$, and b, c, d, $e \to \infty$. It is not hard to see, by means of (4.5), that

$$
{}_8\phi_7[Eq.2.3] \quad \Rightarrow \quad \sum_{i=0}^{\delta+2n-2k} \frac{(q^{-\delta-2n+2k}; q)_i}{(q; q)_i} q^{i\,(\varepsilon+\delta-2-n+k+i)} \tag{4.6a}
$$

$$
= \quad (-1)^{n+k} \frac{(q^{1+\delta}; q)_{2n-2k}}{(q; q)_{n-k}} q^{(n-k)\,(\varepsilon-4)-3\binom{n-k}{2}} \tag{4.6b}
$$

$$
\times \quad \sum_{i=0}^{\max\{\varepsilon+\delta-3,\,2-\varepsilon-2\delta\}} (q^{k-n}; q)_i \frac{(q^{\varepsilon+\delta-2-i}; q)_{\delta+2i}}{(q^{1+\delta}; q)_{2i}} q^{i(3-\varepsilon+n-k+i)} \tag{4.6c}
$$

which results in

$$
\sum_k [Eq.2.3] \quad \Rightarrow \quad \sum_k (-1)^k \frac{(q^{-\delta-2n}; q)_{2k}}{(q; q)_k} q^{k\,(\varepsilon+\delta-n-1)+\binom{k}{2}} \times \{{}_8\phi_7[Eq.2.3]\} \tag{4.7a}
$$

$$
= \quad (-1)^n \frac{(q^{1+\delta}; q)_{2n}}{(q; q)_n} q^{n(\varepsilon-\delta)-5\binom{n+1}{2}} \sum_{k=0}^n \begin{bmatrix} n \\ k \end{bmatrix} q^{k(\delta+k)} \tag{4.7b}
$$

$$
\times \quad \sum_{i=0}^{\max\{\varepsilon+\delta-3,\,2-\varepsilon-2\delta\}} (q^{-k}; q)_i \frac{(q^{\varepsilon+\delta-2-i}; q)_{\delta+2i}}{(q^{1+\delta}; q)_{2i}} q^{i(3-\varepsilon+k+i)} \tag{4.7c}
$$

where the replacement $k \to n - k$ on summation index has been performed from (4.7a) to (4.7b).

On the other hand, we have

$$
\Omega_\delta[Eq.2.3] \quad \Rightarrow \quad \sum_{k=0}^{\delta+2n} \frac{(q^{-\delta-2n}; q)_k}{(q; q)_k} q^{k(\varepsilon-2-3n+2k)} \tag{4.8a}
$$

$$
= \quad (-1)^n q^{n(\varepsilon-\delta)-5\binom{n+1}{2}} \sum_{j=-n}^{\delta+n} (-1)^j \begin{bmatrix} 2n+\delta \\ n+j \end{bmatrix} q^{j\,(\varepsilon-\delta)+5\binom{j}{2}} \tag{4.8b}
$$

where the last line follows from the previous one by changing the summation index $k \to n + j$.

Denote by

$$m = \varepsilon + \delta - 3 = \max\{\varepsilon + \delta - 3,\ 2 - \varepsilon - 2\delta\},\quad (\varepsilon \geq (5 - 3\delta)/2).$$

We may restate $\Omega_\delta[Eq.2.3] = \sum_k [Eq.2.3]$, by interchanging the summation order on (4.7), as

$$\sum_{i=0}^{m} (-1)^i \begin{bmatrix} i + m + \delta \\ 2i + \delta \end{bmatrix} q^{i(2+2\delta-m)+5\binom{i}{2}} \sum_{k=0}^{n-i} \begin{bmatrix} n - i \\ k \end{bmatrix} q^{k(\delta+k+2i)} \qquad (4.9a)$$

$$= \frac{(q;\,q)_n}{(q;\,q)_{\delta+2n}} \sum_{j=-n}^{\delta+n} (-1)^j \begin{bmatrix} 2n + \delta \\ n + j \end{bmatrix} q^{j\,(3-2\delta+m)+5\binom{j}{2}}. \qquad (4.9b)$$

For $\delta = 0,\ 1$ and integral parameters $\mu,\ \nu$, let

$$R_n(\delta, \mu) = \sum_{k=0}^{n-\mu} \begin{bmatrix} n - \mu \\ k \end{bmatrix} q^{k(\delta+k+2\mu)} \qquad (4.10a)$$

$$S_n(\delta, \nu) = \sum_{k=-n}^{\delta+n} (-1)^k \begin{bmatrix} 2n + \delta \\ n + k \end{bmatrix} q^{k(3-2\delta+\nu)+5\binom{k}{2}}. \qquad (4.10b)$$

Then (4.9) may be reformulated further.

$$\sum_{i=0}^{m} (-1)^i \begin{bmatrix} m \\ i \end{bmatrix} (q^{1+\delta+i};\,q)_m\, q^{2i(\delta+i-1/2)+\binom{m-i}{2}} \Big/ \begin{bmatrix} \delta + 2i \\ \delta + i \end{bmatrix} R_n(\delta, i) \qquad (4.11a)$$

$$= q^{\binom{m}{2}} \frac{(q;\,q)_m\,(q;\,q)_n}{(q;\,q)_{\delta+2n}} S_n(\delta, m). \qquad (4.11b)$$

In order to find an expression of R_n in terms of S_n, we recall a useful pair of q-inverse series relations due to Carlitz [6] (see also Chu [7]), which has been used by the author [8] extensively to revisit many q-series identities. For two complex sequences $\{a_k\}$ and $\{b_k\}$, the system of equations

$$f(m) = \sum_{k=0}^{m} (-1)^k \begin{bmatrix} m \\ k \end{bmatrix} q^{\binom{m-k}{2}} \varphi(k; m)\, g(k),\quad (m = 0,\ 1,\ 2,\cdots) \qquad (4.12a)$$

is equivalent to system

$$g(m) = \sum_{k=0}^{m} (-1)^k \begin{bmatrix} m \\ k \end{bmatrix} \frac{a_k + q^k b_k}{\varphi(m;\,k+1)}\, f(k),\quad (m = 0,\ 1,\ 2,\cdots) \qquad (4.12b)$$

where $\varphi(x; n)$ is a polynomial defined by

$$\varphi(x; n) = \prod_{k=0}^{n-1} (a_k + q^x b_k),\quad (n = 0,\ 1,\ 2,\cdots) \qquad (4.12c)$$

with the convention that the empty product is equal to one. Specifying the above inversions by

$$
f(k) = q^{\binom{k}{2}} \frac{(q; q)_k (q; q)_n}{(q; q)_{\delta+2n}} S_n(\delta, k)
$$

$$
g(k) = q^{2k(\delta+k-1/2)} / \left[\begin{array}{c} \delta + 2k \\ \delta + k \end{array} \right] R_n(\delta, k)
$$

$$
\varphi(x; k) = (q^{1+\delta+x}; q)_k
$$

we get the dual relation of (4.11) immediately

$$
\sum_{i=0}^{m} (-1)^i \left[\begin{array}{c} m \\ i \end{array} \right] \frac{1 - q^{1+\delta+2i}}{(q^{1+\delta+m}; q)_{i+1}} q^{\binom{i}{2}} \frac{(q; q)_i (q; q)_n}{(q; q)_{\delta+2n}} S_n(\delta, i) \tag{4.13a}
$$

$$
= q^{2m(\delta+m-1/2)} / \left[\begin{array}{c} \delta + 2m \\ \delta + m \end{array} \right] R_n(\delta, m) \tag{4.13b}
$$

which may be rewritten explicitly as

$$
R_n(\delta, m) = q^{2m(1/2-\delta-m)} \frac{(q; q)_n}{(q; q)_{\delta+2n}} \frac{1 - q^{1+\delta}}{1 - q^{1+\delta+m}} \left[\begin{array}{c} \delta + 2m \\ \delta + m \end{array} \right] \tag{4.14a}
$$

$$
\times \sum_{i=0}^{m} (-1)^i \frac{1 - q^{1+\delta+2i}}{1 - q^{1+\delta}} \left[\begin{array}{c} m \\ i \end{array} \right] \frac{(q; q)_i}{(q^{2+\delta+m}; q)_i} q^{\binom{i}{2}} S_n(\delta, i). \tag{4.14b}
$$

Notice that $S_n(\delta, i)$ may be split as

$$
S_n(\delta, i) = (1 - \delta) \left[\begin{array}{c} \delta + 2n \\ \delta + n \end{array} \right] + \sum_{j=1-\delta}^{n} (-1)^j \left[\begin{array}{c} 2n + \delta \\ n - j \end{array} \right] q^{5\binom{j}{2}}
$$

$$
\times \left\{ q^{2j(1+\delta)-ij} + (-1)^\delta q^{\delta+3j(1+\delta)+i(j+\delta)} \right\}.
$$

Performing the summation with respect to i, we get

$$
[Eq.4.14b] = (1 - \delta) \left[\begin{array}{c} \delta + 2n \\ \delta + n \end{array} \right] \vartheta_m(\delta, 0) + \sum_{j=1-\delta}^{n} (-1)^j \left[\begin{array}{c} 2n + \delta \\ n - j \end{array} \right] q^{5\binom{j}{2}} \tag{4.15a}
$$

$$
\times \left\{ q^{2j(1+\delta)} \vartheta_m(\delta, -j) + (-1)^\delta q^{\delta+3j(1+\delta)} \vartheta_m(\delta, j + \delta) \right\} \tag{4.15b}
$$

where ϑ-function is defined by a terminating bilateral sum

$$
\vartheta_m(\delta, x) = {}_3\psi_3 \left[\begin{array}{ccc} q^{-m}, & q^{(3+\delta)/2}, & -q^{(3+\delta)/2} \\ q^{2+\delta+m}, & q^{(1+\delta)/2}, & -q^{(1+\delta)/2} \end{array} ; q^{x+m} \right]. \tag{4.16}
$$

Shifting the summation variable of ϑ by m, we can reduce (4.15b) as follows

$$[Eq.4.15b] = (-1)^\delta \frac{1 - q^{1+\delta+2m}}{1 - q^{1+\delta}} \begin{bmatrix} q^{-m} \\ q^{2+\delta+m} \end{bmatrix} ; q \Big]_m q^{\delta + m(\delta+m+j) + 3j(1+\delta)} \tag{4.17a}$$

$$\times \left\{ -\delta \frac{1 - q^{-1+\delta}}{1 - q^{-1-\delta-2m}} \begin{bmatrix} q^{-1-\delta-2m} \\ q \end{bmatrix} ; q \Big]_{\delta+m} q^{(m-j)(m+\delta)} \right. \tag{4.17b}$$

$$+ \, {}_3\phi_2 \begin{bmatrix} q^{-1-\delta-2m}, \; q^{-m+(1-\delta)/2}, \; -q^{-m+(1-\delta)/2} \\ q^{-m-(1+\delta)/2}, \; -q^{-m-(1+\delta)/2} \end{bmatrix} ; q^{m-j} \Big] \right\} \tag{4.17c}$$

$$= \frac{q^{m^2}}{1 - q^{1+\delta}} \begin{bmatrix} q^{-m} \\ q^{2+\delta+m} \end{bmatrix} ; q \Big]_m \left\{ q^{2j(1+\delta)-mj} (q^{1-m+j}; q)_{1+\delta+2m} \right. \tag{4.17d}$$

$$\left. + \, (-1)^\delta \, q^{\delta+3j(1+\delta)+m(j+\delta)} (q^{1-\delta-m-j}; q)_{1+\delta+2m} \right\} \tag{4.17e}$$

where $\vartheta_m(0,0) = \delta_{0,m}$ and (4.17b) vanishes on account of $\delta = 0, 1$. Substituting this expression into (4.15), we have after some trivial modification

$$[Eq.4.14\dot{b}] = \frac{q^{m^2}}{1 - q^{1+\delta}} \begin{bmatrix} q^{-m} \\ q^{2+\delta+m} \end{bmatrix} ; q \Big]_m \sum_{k=-n}^{\delta+n} (-1)^k \begin{bmatrix} \delta + 2n \\ k + n \end{bmatrix} q^{5\binom{k}{2}}$$

$$\times \; (q^{1-m-k}; q)_{1+\delta+2m} \; q^{k(3-2\delta+m)}$$

which leads (4.14) to a simplified formula

$$R_n(\delta, m) = (-1)^m q^{-m(1+2\delta)-3\binom{m}{2}} (q; q)_n (q; q)_{\delta+2m} / (q; q)_{\delta+2n} \tag{4.18a}$$

$$\times \sum_{k=-n}^{\delta+n} (-1)^k \begin{bmatrix} \delta + 2n \\ k + n \end{bmatrix} \begin{bmatrix} 1 + \delta + m - k \\ 1 + \delta + 2m \end{bmatrix} q^{k(3-2\delta+m)+5\binom{k}{2}} . \tag{4.18b}$$

Denote the limiting form of $R_n(\delta, \mu)$ by

$$R_{\delta+2\mu} = R_\infty(\delta, \mu) = \sum_{n=0}^{\infty} \frac{q^{n(\delta+n+2\mu)}}{(q; q)_n}. \tag{4.19}$$

Then (4.1a) and (4.1b) may be restated as

$$R_0 = \frac{1}{\prod_{n=0}^{\infty}(1 - q^{1+5n})(1 - q^{4+5n})} \tag{4.20a}$$

$$R_1 = \frac{1}{\prod_{n=0}^{\infty}(1 - q^{2+5n})(1 - q^{3+5n})}. \tag{4.20b}$$

When $n \to \infty$, (4.18) reduces to

$$R_{\delta+2m} = (-1)^m q^{-m(1+2\delta)-3\binom{m}{2}} (q; q)_{\delta+2m} / (q; q)_\infty \tag{4.21a}$$

$$\times \sum_{k=-\infty}^{+\infty} (-1)^k \begin{bmatrix} 1 + \delta + m - k \\ 1 + \delta + 2m \end{bmatrix} q^{k(3-2\delta+m)+5\binom{k}{2}} . \tag{4.21b}$$

Replacing the Gaussian binomial coefficient by

$$\begin{bmatrix} 1+\delta+m-k \\ 1+\delta+2m \end{bmatrix} = \sum_i (-1)^i \begin{bmatrix} 1+\delta+2m \\ i \end{bmatrix} q^{i(1-m-k)+\binom{i}{2}} / (q;q)_{1+\delta+2m}$$

interchanging the summation order and then using the Jacobi product identity (4.2), we have

$$R_{\delta+2m} = (-1)^m \, q^{-m(1+2\delta)-3\binom{m}{2}} / (1-q^{1+\delta+2m}) \tag{4.21c}$$

$$\times \sum_i (-1)^i \begin{bmatrix} 1+2\delta+m \\ i \end{bmatrix} \frac{[q^5,\, q^{2+2\delta-m+i},\, q^{3-2\delta+m-i};\, q^5]_\infty}{(q;q)_\infty} q^{i(1-m)+\binom{i}{2}}. \tag{4.21d}$$

Rename the summation index according to

$$2\delta - m + i = \gamma + 5k, \qquad (\gamma = 0,\, \pm 1,\, \pm 2). \tag{4.22a}$$

Then

$$[q^5,\, q^{2+\delta-m+i},\, q^{3-2\delta+m-i};\, q^5]_\infty \tag{4.22b}$$

$$\Rightarrow \quad \frac{(q^{3-\gamma-5k};\, q^5)_k}{(q^{2+\gamma};\, q^5)_k} [q^5,\, q^{2+\gamma},\, q^{3-\gamma};\, q^5]_\infty \tag{4.22c}$$

$$= \quad (-1)^k q^{k(1-2\gamma-5k)/2} [q^5,\, q^{2+\gamma},\, q^{3-\gamma};\, q^5]_\infty. \tag{4.22d}$$

Substituting these relations into (4.21), we find

$$R_{\delta+2m} \quad = \quad \frac{q^{(\delta+m)(1-2m)}}{1-q^{1+\delta+2m}} \sum_k \sum_{\gamma=-2}^2 (-1)^\gamma \begin{bmatrix} 1+\delta+2m \\ m+5k+\gamma-2\delta \end{bmatrix} \tag{4.23a}$$

$$\times \quad \frac{[q^5,\, q^{2+\gamma},\, q^{3-\gamma};\, q^5]_\infty}{(q;q)_\infty} q^{(\gamma+5k)(1+2k-2\delta)+(\gamma+4k)(\gamma-1)/2} \tag{4.23b}$$

where the exponent of q has been calculated as

$$\binom{\gamma+5k+m-2\delta}{2} + k(1-2\gamma-5k)/2$$

$$+ (\gamma+5k+m-2\delta)(1-m) - m(1+2\delta) - 3\binom{m}{2}$$

$$= \binom{m-2\delta}{2} + (m-2\delta)(1-m) - m(1+2\delta) - 3\binom{m}{2}$$

$$+ (\gamma+5k)(1-2\delta) + \binom{\gamma+5k}{2} + k(1-2\gamma-5k)/2$$

$$= (\delta+m)(1-2m) + (\gamma+5k)(1+2k-2\delta) + (\gamma+4k)(\gamma-1)/2$$

in view of

$$\binom{u+v}{2} = \binom{u}{2} + uv + \binom{v}{2}.$$

Noting that

$$\frac{[q^5, q^{2+\gamma}, q^{3-\gamma}; q^5]_\infty}{(q; q)_\infty} = \begin{cases} R_0, & \gamma = 0, 1 \\ R_1, & \gamma = 2, -1 \\ 0, & \gamma = -2 \end{cases}$$

we may regroup the terms of (4.23) as follows

$$R_{\delta+2m} = q^{(\delta+m)(1-2m)}/(1 - q^{1+\delta+2m}) \tag{4.24a}$$

$$\times \left\{ R_0 \sum_k \begin{bmatrix} 1 + \delta + 2m \\ m + 5k - 2\delta \end{bmatrix} q^{5k(1+2k-2\delta)-2k} \right. \tag{4.24b}$$

$$-R_0 \sum_k \begin{bmatrix} 1 + \delta + 2m \\ m + 5k + 1 - 2\delta \end{bmatrix} q^{(1+5k)(1+2k-2\delta)} \tag{4.24c}$$

$$+R_1 \sum_k \begin{bmatrix} 1 + \delta + 2m \\ m + 5k + 2 - 2\delta \end{bmatrix} q^{(2+5k)(1+2k-2\delta)+(1+2k)} \tag{4.24d}$$

$$\left. -R_1 \sum_k \begin{bmatrix} 1 + \delta + 2m \\ m + 5k - 1 - 2\delta \end{bmatrix} q^{(-1+5k)(1+2k-2\delta)+(1-4k)} \right\}. \tag{4.24e}$$

Replacing the summation variable k by $\delta - k$ in (4.24c) and (4.24e), we get the simplified expression of R_n in terms of R_0 and R_1.

$$R_{\delta+2m} = q^{(\delta+m)(1-2m)} R_0 \sum_k \begin{bmatrix} 1 + \delta + 2m \\ m + 5k - 2\delta \end{bmatrix} \frac{q^{1+5\delta-10k}}{1 - q^{1+\delta+2m}} q^{3k+10k(k-\delta)} \tag{4.25a}$$

$$- q^{(\delta+m)(1-2m)} R_1 \sum_k \begin{bmatrix} 1 + \delta + 2m \\ m + 5k - 1 - 2\delta \end{bmatrix} \frac{q^{3+5\delta-10k}}{1 - q^{1+\delta+2m}} q^{2\delta-k+10k(k-\delta)} \tag{4.25b}$$

From the last formula, we can compute the sums of Rogers-Ramanujan type. A short list is displayed as follows:

$$\delta = 0, \ m = 0 : \quad R_0 = R_0$$
$$\delta = 1, \ m = 0 : \quad R_1 = R_1$$
$$\delta = 0, \ m = 1 : \quad R_2 = q^{-1}(R_0 - R_1)$$
$$\delta = 1, \ m = 1 : \quad R_3 = -q^{-3}\{R_0 - (1+q)R_1\}$$
$$\delta = 0, \ m = 2 : \quad R_4 = q^{-6}\{(1+q^2)R_0 - (1+q+q^2)R_1\}$$
$$\delta = 1, \ m = 2 : \quad R_5 = -q^{-10}\{(1+q^2+q^3)R_0 - (1+q+q^2+q^3+q^4)R_1\}$$

$\cdots \ \cdots \qquad\qquad \cdots \ \cdots \ \cdots$

5 The q-Clausen-Orr-Formulae

Throughout this section, the basic hypergeometric series will be denoted, in order to specify the base q, by $_{1+p}\phi_p \left[\begin{matrix} a_0, & a_1, & \cdots, & a_p \\ & b_1, & \cdots, & b_p \end{matrix} ; q, z \right]$ with reference to (0.1).

Following the approach of Jackson [14], consider the product of two basic hypergeometric series

$$\sum_n A_n \, z^n \; = \; {}_2\phi_1 \left[\begin{matrix} a^2 q^\alpha, \, b^2 q^\beta \\ a^2 b^2 q^\gamma \end{matrix} ; q^2, \, z \right] \times {}_2\phi_1 \left[\begin{matrix} a^2 q^{\alpha'}, \, b^2 q^{\beta'} \\ a^2 b^2 q^{\gamma'} \end{matrix} ; q^2, \, qz \right] \quad (5.1a)$$

where

$$A_n \; = \; \sum_k \left[\begin{matrix} a^2 q^\alpha, \, b^2 q^\beta \\ q^2, \, a^2 b^2 q^\gamma \end{matrix} ; q^2 \right]_k \left[\begin{matrix} a^2 q^{\alpha'}, \, b^2 q^{\beta'} \\ q^2, \, a^2 b^2 q^{\gamma'} \end{matrix} ; q^2 \right]_{n-k} q^{n-k} \quad (5.1b)$$

$$= \; \left[\begin{matrix} a^2 q^\alpha, & b^2 q^\beta \\ q^2, & a^2 b^2 q^\gamma \end{matrix} ; q^2 \right]_n \times$$

$$\times \, {}_4\phi_3 \left[\begin{matrix} q^{-2n}, \, a^2 q^{\alpha'}, & b^2 q^{\beta'}, & q^{2-\gamma-2n}/a^2 b^2 \\ q^{2-\alpha-2n}/a^2, & q^{2-\beta-2n}/b^2, & a^2 b^2 q^{\gamma'} \end{matrix} ; q^2, \, q^{3-\alpha-\beta+\gamma} \right] \quad (5.1c)$$

$$= \; q^n \left[\begin{matrix} a^2 q^{\alpha'}, & b^2 q^{\beta'} \\ q^2, & a^2 b^2 q^{\gamma'} \end{matrix} ; q^2 \right]_n \times$$

$$\times \, {}_4\phi_3 \left[\begin{matrix} q^{-2n}, \, a^2 q^\alpha, & b^2 q^\beta, & q^{2-\gamma'-2n}/a^2 b^2 \\ q^{2-\alpha'-2n}/a^2, \, q^{2-\beta'-2n}/b^2, & a^2 b^2 q^\gamma \end{matrix} ; q^2, \, q^{1-\alpha'-\beta'+\gamma'} \right]. \quad (5.1d)$$

When A_n has a closed form in terms of q-factorial fraction, the q-series product formulae (including the q-Clausen formula) may be established consequently.

For an alternative q-analogue of the Clausen theorem and other proofs of Jackson's q-Clausen formula, refer to [11] and the articles cited there.

Example 5.1 (Jackson's q-Clausen formula) *For*

$$\alpha = 0, \quad \beta = 0 \quad \gamma = 1$$
$$\alpha' = 0, \quad \beta' = 0 \quad \gamma' = 1$$

the terminating series (5.1c) and (5.1d) may be evaluated as

$$_4\phi_3 \left[\begin{matrix} q^{-n}, & A, & B, & q^{1/2-n}/AB \\ & q^{1-n}/A, & q^{1-n}/B, & q^{1/2}AB \end{matrix} ; q^\varepsilon \right] \tag{5.2a}$$

$$\underset{\varepsilon=1,2}{=\!=\!=} q^{n(\varepsilon-2)/2} \left[\begin{matrix} q, \ AB \\ A, \ B \end{matrix} ; q \right]_n \left[\begin{matrix} A, \ B \\ q^{1/2}, \ AB \end{matrix} ; q^{1/2} \right]_n \tag{5.2b}$$

by means of Examples 2.1 *and* Example 2.11 *specified, according the parity of n, as follows*

$$_5\phi_4 \left[\begin{matrix} q^{-2m}, & A, & B, & q^{1/2-m}, & q^{1/2-2m}/AB \\ & q^{1-2m}/A, & q^{1-2m}/B, & q^{1/2-m}, & q^{1/2}AB \end{matrix} ; q^\varepsilon \right]$$

$$\underset{\varepsilon=1,2}{=\!=\!=} q^{m(\varepsilon-2)} \left[\begin{matrix} Aq^{1/2}, & Bq^{1/2}, & q^{m+1}, & ABq^m \\ q^{1/2}, & ABq^{1/2}, & Aq^m, & Bq^m \end{matrix} ; q \right]_m$$

$$_5\phi_4 \left[\begin{matrix} q^{-1-2m}, & A, & B, & q^{1/2-m}, & q^{-1/2-2m}/AB \\ & q^{-2m}/A, & q^{-2m}/B, & q^{1/2-m}, & q^{1/2}AB \end{matrix} ; q^\varepsilon \right]$$

$$\underset{\varepsilon=1,2}{=\!=\!=} q^{(m+\frac{1}{2})(\varepsilon-2)} (1+q^{m+\frac{1}{2}}) \left[\begin{matrix} Aq^{1/2}, & Bq^{1/2}, q^{m+1}, & ABq^{m+1} \\ q^{1/2}, & ABq^{1/2}, Aq^{m+1}, & Bq^{m+1} \end{matrix} ; q \right]_m .$$

The corresponding product (5.1a) yields Jackson's [14] *formula*

$$_2\phi_1 \left[\begin{matrix} a^2, \ b^2 \\ qa^2b^2 \end{matrix} ; q^2, \ z \right] \times {}_2\phi_1 \left[\begin{matrix} a^2, \ b^2 \\ qa^2b^2 \end{matrix} ; q^2, \ qz \right] \tag{5.3a}$$

$$= {}_4\phi_3 \left[\begin{matrix} a^2, & b^2, & ab, & -ab \\ & a^2b^2, & abq^{1/2}, & -abq^{1/2} \end{matrix} ; q, \ z \right] \tag{5.3b}$$

which is a q-analogue of the Clausen (1828) *theorem:*

$$_2F_1 \left[\begin{matrix} A, \ B \\ A+B+1/2 \end{matrix} ; z \right] \times {}_2F_1 \left[\begin{matrix} A, \ B \\ A+B+1/2 \end{matrix} ; z \right] \tag{5.4a}$$

$$= {}_3F_2 \left[\begin{matrix} 2A, & 2B, & A+B \\ & 2A+2B, & A+B+1/2 \end{matrix} ; z \right] . \tag{5.4b}$$

Example 5.2 (the q-Orr formula) *For*

$$\alpha = 0, \quad \beta = 0 \quad \gamma = -1$$
$$\alpha' = 0, \quad \beta' = 0 \quad \gamma' = 1$$

the terminating series (5.1c) may be evaluated as

$$_4\phi_3 \left[\begin{array}{cccc} q^{-n}, & A, & B, & q^{3/2-n}/AB \\ & q^{1-n}/A, & q^{1-n}/B, & q^{1/2}AB \end{array} ; q \right] \tag{5.5a}$$

$$= \left[\begin{array}{c} q, AB, ABq^{-1/2} \\ A, \ B, \ ABq^{1/2} \end{array} ; q \right]_n \left[\begin{array}{c} A, \ B \\ q^{1/2}, \ q^{-1/2}AB \end{array} ; q^{1/2} \right]_n \tag{5.5b}$$

by means of Examples 2.3 *and* Example 2.13 *specified, according the parity of n, as follows*

$$_5\phi_4 \left[\begin{array}{ccccc} q^{-2m}, & A, & B, & q^{1/2-m}, & q^{3/2-2m}/AB \\ & q^{1-2m}/A, & q^{1-2m}/B, & q^{1/2-m}, & q^{1/2}AB \end{array} ; q \right]$$

$$= \left[\begin{array}{ccccc} Aq^{1/2}, & Bq^{1/2}, & q^{m+1}, & ABq^m, & ABq^{m-1/2} \\ q^{1/2}, & ABq^{1/2}, & Aq^m, & Bq^m, & ABq^{m+1/2} \end{array} ; q \right]_m$$

$$_5\phi_4 \left[\begin{array}{ccccc} q^{-1-2m}, & A, & B, & q^{1/2-m}, & q^{1/2-2m}/AB \\ & q^{-2m}/A, & q^{-2m}/B, & q^{1/2-m}, & q^{1/2}AB \end{array} ; q \right]$$

$$= \left[\begin{array}{c} q^{2m+1}, ABq^m \\ q^{m+1/2}, ABq^{2m+1/2} \end{array} \right] \left[\begin{array}{c} Aq^{1/2}, Bq^{1/2}, q^{m+1}, ABq^{m+1} \\ q^{1/2}, ABq^{1/2}, Aq^{m+1}, Bq^{m+1} \end{array} ; q \right]_m.$$

The terminating series (5.1d) is the reversal of summation (5.5a)

$$_4\phi_3 \left[\begin{array}{cccc} q^{-n}, & A, & B, & q^{1/2-n}/AB \\ & q^{1-n}/A, & q^{1-n}/B, & q^{-1/2}AB \end{array} ; q \right] \tag{5.5c}$$

$$= q^{-n/2} \left[\begin{array}{c} q, AB \\ A, \ B \end{array} ; q \right]_n \left[\begin{array}{c} A, \ B \\ q^{1/2}, \ q^{-1/2}AB \end{array} ; q^{1/2} \right]_n \tag{5.5d}$$

which may be derived from Examples 2.4 *and* Example 2.10 *in the same way.*

The corresponding product (5.1a) yields the formula

$$_2\phi_1 \left[\begin{array}{c} a^2, b^2 \\ q^{-1}a^2b^2 \end{array} ; q^2, z \right] \times _2\phi_1 \left[\begin{array}{c} a^2, b^2 \\ qa^2b^2 \end{array} ; q^2, qz \right] \tag{5.6a}$$

$$= _4\phi_3 \left[\begin{array}{cccc} a^2, & b^2, & ab, & -ab \\ & q^{-1}a^2b^2, & abq^{1/2}, & -abq^{1/2} \end{array} ; q, z \right] \tag{5.6b}$$

which is a q-analogue of the Orr (1899) *theorem.*

$$_2F_1 \left[\begin{array}{c} A, B \\ A+B-1/2 \end{array} ; z \right] \times _2F_1 \left[\begin{array}{c} A, B \\ A+B+1/2 \end{array} ; z \right] \tag{5.7a}$$

$$= _3F_2 \left[\begin{array}{cc} 2A, 2B, & A+B \\ 2A+2B-1, & A+B+1/2 \end{array} ; z \right]. \tag{5.7b}$$

Example 5.3 (the q-Orr formula) *For*

$$\alpha = 0, \quad \beta = 1 \quad \gamma = 0$$
$$\alpha' = 0, \quad \beta' = -1 \quad \gamma' = 0$$

the terminating series (5.1c) may be evaluated as

$$
{}_4\phi_3 \left[\begin{array}{cccc} q^{-n}, & A, & Bq^{1/2}, & q^{1-n}/AB \\ & q^{1-n}/A, & q^{3/2-n}/B, & AB \end{array} ; q \right] \tag{5.8a}
$$

$$
= q^{-n/2} \left[\begin{array}{c} q, \ ABq^{-1/2} \\ A, \ Bq^{-1/2} \end{array} ; q \right]_n \left[\begin{array}{c} A, \ B \\ q^{1/2}, \ q^{-1/2}AB \end{array} ; q^{1/2} \right]_n \tag{5.8b}
$$

by means of Examples 2.3 *and* Example 2.13 *specified, according the parity of* n, *as follows*

$$
{}_5\phi_4 \left[\begin{array}{ccccc} q^{-2m}, & A, & Bq^{1/2}, & q^{1/2-m}, & q^{1-2m}/AB \\ & q^{1-2m}/A, & q^{3/2-2m}/B, & q^{1/2-m}, & AB \end{array} ; q \right]
$$

$$
= q^{-m} \left[\begin{array}{ccccc} Aq^{1/2}, & B, & q^{m+1}, & ABq^{m-1/2}, & Bq^{1/2} \\ q^{1/2}, & AB, & Aq^m, & Bq^{m-1/2}, & Bq^{-1/2} \end{array} ; q \right]_m
$$

$$
{}_5\phi_4 \left[\begin{array}{ccccc} q^{-1-2m}, & A, & Bq^{1/2}, & q^{1/2-m}, & q^{-2m}/AB \\ & q^{-2m}/A, & q^{1/2-2m}/B, & q^{1/2-m}, & AB \end{array} ; q \right]
$$

$$
= q^{-m-1/2} \left[\begin{array}{c} q^{2m+1}, \ Bq^m \\ q^{m+1/2}, \ Bq^{-1/2} \end{array} \right] \left[\begin{array}{c} Aq^{1/2}, \ B, q^{m+1}, \ ABq^{m+1/2} \\ q^{1/2}, \ AB, Aq^{m+1}, \ Bq^{m+1/2} \end{array} ; q \right]_m .
$$

The terminating series (5.1d) is the reversal of summation (5.8a)

$$
{}_4\phi_3 \left[\begin{array}{cccc} q^{-n}, & A, & Bq^{-1/2}, & q^{1-n}/AB \\ & q^{1-n}/A, & q^{1/2-n}/B, & AB \end{array} ; q \right] \tag{5.8c}
$$

$$
= \left[\begin{array}{c} q, \ ABq^{-1/2} \\ A, \ Bq^{1/2} \end{array} ; q \right]_n \left[\begin{array}{c} A, \ B \\ q^{1/2}, \ q^{-1/2}AB \end{array} ; q^{1/2} \right]_n \tag{5.8d}
$$

which may be derived from Examples 2.4 *and* Example 2.10 *in the same way.*

The corresponding product (5.1a) yields the formula

$$
{}_2\phi_1 \left[\begin{array}{c} a^2, \ b^2 q \\ a^2 b^2 \end{array} ; q^2, z \right] \times {}_2\phi_1 \left[\begin{array}{c} a^2, \ b^2/q \\ a^2 b^2 \end{array} ; q^2, qz \right] \tag{5.9a}
$$

$$
= {}_4\phi_3 \left[\begin{array}{cccc} a^2, & b^2, & abq^{-1/2}, & -abq^{-1/2} \\ & q^{-1}a^2 b^2, & ab, & -ab \end{array} ; q, z \right] \tag{5.9b}
$$

which is a q-analogue of the Orr (1899) theorem.

$$
{}_2F_1 \left[\begin{array}{c} A, \ B+1/2 \\ A+B \end{array} ; z \right] \times {}_2F_1 \left[\begin{array}{c} A, \ B-1/2 \\ A+B \end{array} ; z \right] \tag{5.10a}
$$

$$
= {}_3F_2 \left[\begin{array}{ccc} 2A, & 2B, & A+B-1/2 \\ 2A+2B-1, & & A+B \end{array} ; z \right] . \tag{5.10b}
$$

Example 5.4 (A q-series product formula) *For*

$$\alpha = 1, \quad \beta = 1 \quad \gamma = 1$$
$$\alpha' = -1, \quad \beta' = -1 \quad \gamma' = -1$$

the terminating series (5.1c) may be evaluated as

$$
{}_4\phi_3\left[\begin{array}{cccc} q^{-n}, & Aq^{1/2}, & Bq^{1/2}, & q^{3/2-n}/AB \\ & q^{3/2-n}/A, & q^{3/2-n}/B, & ABq^{1/2} \end{array} ; q \right] \quad (5.11a)
$$

$$
= q^{-n/2}\left[\begin{array}{c} q, \, AB, \quad ABq^{-1/2} \\ Aq^{-1/2}, \, Bq^{-1/2}, \, ABq^{1/2} \end{array} ; q \right]_n \left[\begin{array}{c} A, \, B \\ q^{1/2}, \, q^{-1/2}AB \end{array} ; q^{1/2} \right]_n \quad (5.11b)
$$

by means of Examples 2.28 *and* Example 2.33 *specified, according the parity of n, as follows*

$$
{}_5\phi_4\left[\begin{array}{ccccc} q^{-2m}, & Aq^{1/2}, & Bq^{1/2}, & q^{1/2-m}, & q^{3/2-2m}/AB \\ & q^{3/2-2m}/A, & q^{3/2-2m}/B, & q^{1/2-m}, & ABq^{1/2} \end{array} ; q \right]
$$

$$
= \quad q^{-m}\left[\begin{array}{ccc} Aq^{1/2}, & B^{1/2}, & ABq^{m-1/2} \\ Aq^{-1/2}, & Bq^{-1/2}, & ABq^{m+1/2} \end{array} ; q \right]_m
$$

$$
\times \left[\begin{array}{cccc} A, & B, & q^{m+1}, & ABq^m \\ q^{1/2}, & AB^{1/2}, & Aq^{m-1/2}, & Bq^{m-1/2} \end{array} ; q \right]_m
$$

$$
{}_5\phi_4\left[\begin{array}{ccccc} q^{-1-2m}, & Aq^{1/2}, & Bq^{1/2}, & q^{1/2-m}, & q^{1/2-2m}/AB \\ & q^{1/2-2m}/A, & q^{1/2-2m}/B, & q^{1/2-m}, & ABq^{1/2} \end{array} ; q \right]
$$

$$
= \quad q^{-(m+1/2)}\left[\begin{array}{c} Aq^{m-1/2}, Bq^{m-1/2}, ABq^{m+1/2} \\ Aq^{-1/2}, \quad Bq^{-1/2}, \quad ABq^{2m+1/2} \end{array} \right]
$$

$$
\times \left[\begin{array}{cccc} A, & B, & q^{m+1}, & ABq^m \\ q^{1/2}, & ABq^{1/2}, & Aq^{m-1/2}, & Bq^{m-1/2} \end{array} ; q \right]_{m+1} .
$$

The terminating series (5.1d) is the reversal of summation (5.11a)

$$
{}_4\phi_3\left[\begin{array}{cccc} q^{-n}, & Aq^{-1/2}, & Bq^{-1/2}, & q^{1/2-n}/AB \\ & q^{1/2-n}/A, & q^{1/2-n}/B, & ABq^{-1/2} \end{array} ; q \right] \quad (5.11c)
$$

$$
= \left[\begin{array}{c} q, \, AB \\ Aq^{1/2}, \, Bq^{1/2} \end{array} ; q \right]_n \left[\begin{array}{c} A, \, B \\ q^{1/2}, \, q^{-1/2}AB \end{array} ; q^{1/2} \right]_n \quad (5.11d)
$$

which may be derived from Examples 2.20 *and* Example 2.22 *in the same way.*

The corresponding product (5.1a) yields a formula

$$
{}_2\phi_1\left[\begin{array}{c} a^2q, \, b^2q \\ a^2b^2q \end{array} ; q^2, z \right] \times {}_2\phi_1\left[\begin{array}{c} a^2/q, \, b^2/q \\ a^2b^2/q \end{array} ; q^2, qz \right] \quad (5.12a)
$$

$$
= {}_4\phi_3\left[\begin{array}{cccc} a^2, & b^2, & ab, & -ab \\ & q^{-1}a^2b^2, & abq^{1/2}, & -abq^{1/2} \end{array} ; q, z \right] \quad (5.12b)
$$

which is a q-analogue of the following hypergeometric product

$$_2F_1 \left[\begin{array}{c} A+1/2,\ B+1/2 \\ A+B+1/2 \end{array} ; z \right] \times {}_2F_1 \left[\begin{array}{c} A-1/2,\ B-1/2 \\ A+B-1/2 \end{array} ; z \right] \quad (5.13a)$$

$$= {}_3F_2 \left[\begin{array}{cc} 2A,\quad 2B, & A+B \\ 2A+2B-1, & A+B+1/2 \end{array} ; z \right]. \quad (5.13b)$$

They may have already appeared in the vast literature on the subject.

Example 5.5 (Another q-series product formula) *Consider the formal product*

$$\sum_n B_n\, z^n \ = \ {}_2\phi_1 \left[\begin{array}{c} q^{1/2}a^2,\quad q^{1/2}b^2 \\ q^2 a^2 b^2 \end{array} ; q^2,\ z \right] \quad (5.14a)$$

$$\times\quad {}_2\phi_1 \left[\begin{array}{c} q^{1/2}/a^2,\ q^{1/2}/b^2 \\ q^2/a^2 b^2 \end{array} ; q^2,\ qz \right] \quad (5.14b)$$

where

$$B_n \ = \ \left[\begin{array}{cc} q^{1/2}a^2,\ q^{1/2}b^2 \\ q^2,\quad q^2 a^2 b^2 \end{array} ; q^2 \right]_n \ \times$$

$$\times\, {}_4\phi_3 \left[\begin{array}{cccc} q^{-2n}/a^2 b^2,\ q^{1/2}/a^2, & q^{1/2}/b^2, & q^{-2n} \\ q^{3/2-2n}/a^2, & q^{3/2-2n}/b^2, & q^2/a^2 b^2 \end{array} ; q^2,\ q^4 \right]$$

$$= \ q^n \left[\begin{array}{c} q^{1/2}/a^2,\ q^{1/2}/b^2 \\ q^2,\quad q^2/a^2 b^2 \end{array} ; q^2 \right]_n \ \times$$

$$\times\, {}_4\phi_3 \left[\begin{array}{cccc} q^{-2n}a^2 b^2,\ q^{1/2}a^2, & q^{1/2}b^2, & q^{-2n} \\ q^{3/2-2n}a^2, & q^{3/2-2n}b^2, & q^2 a^2 b^2 \end{array} ; q^2,\ q^2 \right]$$

which can be evaluated as

$$B_n \ = \ \frac{1+ab}{2} \left[\begin{array}{ccccc} q^{1/2}, & -q^{1/2}, & q^{1/2}a/b, & q^{1/2}b/a \\ q, & -q, & qab, & q/ab \end{array} ; q \right]_n$$

$$+\ \frac{1-ab}{2} \left[\begin{array}{ccccc} q^{1/2}, -q^{1/2}, & -q^{1/2}a/b, -q^{1/2}b/a \\ q, & -q, & -qab, & -q/ab \end{array} ; q \right]_n$$

by means of a well-poised summation formula due to Jain & Verma [15, Eqs.1.3 & 4.4]

$$_4\phi_3 \left[\begin{array}{cccc} ABq^{-m}, & q^{1/4}A, & q^{1/4}B, & q^{-m} \\ Aq^{3/4-m}, & Bq^{3/4-m}, & qAB \end{array} ; q^\varepsilon \right] \quad (\varepsilon = 1,\ 2) \quad (5.15a)$$

$$= \frac{q^{m(\varepsilon-2)/2}}{2} \left[\begin{array}{cc} q^{1/2}, & q/AB \\ q^{1/4}/A, & q^{1/4}/B \end{array} ; q \right]_m \quad (5.15b)$$

$$\times \left\{ (1+\sqrt{(AB)^{3-\varepsilon}}) \left[\begin{array}{cc} q^{1/4}\sqrt{A/B}, & q^{1/4}\sqrt{B/A} \\ q^{1/2}\sqrt{AB}, & q^{1/2}/\sqrt{AB} \end{array} ; q^{1/2} \right]_m \quad (5.15c)$$

$$+ (1-\sqrt{(AB)^{3-\varepsilon}}) \left[\begin{array}{cc} -q^{1/4}\sqrt{A/B}, & -q^{1/4}\sqrt{B/A} \\ -q^{1/2}\sqrt{AB}, & -q^{1/2}/\sqrt{AB} \end{array} ; q^{1/2} \right]_m \right\}. \quad (5.15d)$$

The corresponding product (5.14) yields

$$
{}_2\phi_1 \left[\begin{matrix} q^{1/2}a^2,\ q^{1/2}b^2 \\ q^2 a^2 b^2 \end{matrix}\ ;\ q^2,\ z \right] \times {}_2\phi_1 \left[\begin{matrix} q^{1/2}/a^2,\ q^{1/2}/b^2 \\ q^2/a^2 b^2 \end{matrix}\ ;\ q^2,\ qz \right] \tag{5.16a}
$$

$$
= \frac{1+ab}{2}\ {}_4\phi_3 \left[\begin{matrix} q^{1/2},\ -q^{1/2},\ q^{1/2}a/b,\ q^{1/2}b/a \\ -q,\ qab,\ q/ab \end{matrix}\ ;\ q,\ z \right] \tag{5.16b}
$$

$$
+ \frac{1-ab}{2}\ {}_4\phi_3 \left[\begin{matrix} q^{1/2},\ -q^{1/2},\ -q^{1/2}a/b,\ -q^{1/2}b/a \\ -q,\ -qab,\ -q/ab \end{matrix}\ ;\ q,\ z \right] \tag{5.16c}
$$

which is a q-analogue of the following hypergeometric product (cf. [1, p.100]) &
[12, Ex.4.92])

$$
{}_2F_1 \left[\begin{matrix} 1/4+A,\ 1/4+B \\ 1+A+B \end{matrix}\ ;\ z \right] \times {}_2F_1 \left[\begin{matrix} 1/4-A,\ 1/4-B \\ 1-A-B \end{matrix}\ ;\ z \right] \tag{5.17a}
$$

$$
= {}_3F_2 \left[\begin{matrix} 1/2,\ 1/2+A-B,\ 1/2-A+B \\ 1+A+B,\ 1-A-B \end{matrix}\ ;\ z \right]. \tag{5.17b}
$$

Of course, a more "natural" q-analogue of this product formula is desirable, which
the author is unfortunately unable to establish at present.

APPENDIX: Mathematica Program

for Almost Poised Basic Series

In[1]:=

(* Program for Extra-factor $\chi_\delta \begin{bmatrix} \alpha, \gamma; \varepsilon \\ b, c; n \end{bmatrix} \{\lambda, \mu, \nu\} \end{bmatrix}$ of Almost-Poised-Series *)

(* Factorial-function *)

```
f[x_,k_] := Product[(1-x*q^{Sign[k]*(i-Ceiling[Sign[k]/2])})
                    ^{Sign[k]},{i,1,k*Sign[k]}]
```

(* Factorial-fraction *)

```
ff[k_,{a_,b_,c_,r_,s_,t_},{u_,v_,w_,x_,y_,z_}] :=
    (f[a,k]*f[b,k]*f[c,k]*f[r,k]*f[s,k]*f[t,k])/
    (f[u,k]*f[v,k]*f[w,k]*f[x,k]*f[y,k]*f[z,k])
```

(* Inner-basic-series *)

```
h[k_,{e_,d_},{a_,r_},{l_,u_,v_}] :=Sum[q^{j*(e+l-u-v+k)}*
        ff[j,{q^{l-1-2*k},q^{-d-n-k},b*q^{n-k},
            c*q^{n-k},q^{u-n-k}/b,q^{v-n-k}/c},
            {q,q^{l-1-n-k},b*q^{l-u+n-k},c*q^{l-v+n-k},
            q^{a-n-k}/b,q^{r-n-k}/c}],{j,0,1+2*k-1}]
```

(* Summation *)

```
summ[{e_,d_},{a_,r_},{l_,u_,v_}] :=
    Sum [ff[k,{q^{-n},q^{2-l+n},q^{1+u-l-n}/b,
            q^{1+v-l-n}/c,b*q^{1-a+n},c*q^{1-r+n}},
            {0, q^{1+d+n},q^{1-n}/b,
            q^{1-n}/c,b*c*q^{1+l-u-v+n},0}]*
        (-1)^{k}*h[k,{e,d},{a,r},{l,u,v}]*
        q^{k*((3+k)/2-e+d-u-v+l+a+r)}/(f[q^{2-l},2*k]),
        {k,Floor[l/2],Max[-e+u+v-l,e-d-1+u+v-l-a-r]}]
```

(* major-part for almost-poised-series *)

```
p[a_,r_]:=((a+r)+(a-r)*Sign[a-r])/2
q[l_,u_,v_]:=u+v-l

mpart[n_,{e_,d_},{a_,r_}] :=(f[q^(-d-2*n), n]*
                f[q^(p[a,r]-2*n)/(b*c),n])/
        (f[q^(a-2*n)/b, n]*f[q^(r-2*n)/c, n])
```

93

```
(* modified-factor *)

    mf[{a_,r_},{l_,u_,v_}] :=
              f[q^(p[a,r]-n)/(b*c),q[l,u,v]-p[a,r]]/
              f[q^(p[a,r]-2*n)/(b*c),q[l,u,v]-p[a,r]]


    nf[e_,{l_,u_,v_}] := q^{n(e+2*l-u-v-1)}*
                       (f[q^(1+n),1-l]*f[b,1-u]*f[c,1-v])/
                       (f[q,1-l]*f[b*q^n,1-u]*f[c*q^n,1-v])

    mfct[e_,{a_,r_},{l_,u_,v_}] :=
          Times[mf[{a,r},{l,u,v}],nf[e,{l,u,v}]]

(* Extra-factor *)

    extra[{e_,d_},{a_,r_},{l_,u_,v_}] :=
        Cancel[Times[mfct[e,{a,r},{l,u,v}],
               summ[{e,d},{a,r},{l,u,v}]]]


(* definition for almost-poised-series *)

    aps[n_,{e_,d_},{a_,r_}] :=
        Sum[{q^(e-n)/(b*c)}^j*ff[j,{q^{-d-2*n},b,c,0,0,0},
           {q,q^{a-2*n}/b,q^{r-2*n}/c,0,0,0}],{j,0,d+2*n}]

(* verification *)

    check[n_,{e_,d_},{a_,r_}] :=
          aps[n,{e,d},{a,r}]/mpart[n,{e,d},{a,r}]

(* check[n,{e,d},{a,r}]===extra[{e,d},{a,r},{l,u,v}]*)


(* new-block *)

    extra[{e,d},{a,r},{l,u,v}]
```

(*demonstration for $\chi_0 \begin{bmatrix} \alpha, \gamma; \varepsilon \\ b, c; n \end{bmatrix} \{\lambda, \mu, \nu\}$ *)

```
In[2]:=
  extra[{1,0},{1,1},{1,1,1}]
Out[2]=
  {1}
```

```
In[3]:=
   extra[{2,0},{1,1},{1,1,1}]
Out[3]=
     n
   {q }

In[4]:=
   extra[{2,0},{1,2},{1,1,2}]
Out[4]=
           n
    q - c q
   {--------}
    -c + q

In[5]:=
   extra[{1,0},{0,1},{1,1,1}]
Out[5]=
   {1}

In[6]:=
   extra[{2,0},{2,2},{1,2,2}]
Out[6]=
         n         n
   (q - b q ) (q - c q )
   {--------------------}
     n
    q  (-b + q) (-c + q)

In[7]:=
   extra[{3,0},{2,2},{1,2,2}]
Out[7]=
    n         n         n
   q  (q - b q ) (q - c q )
   {-----------------------}
       (-b + q) (-c + q)

In[8]:=
   extra[{0,0},{0,0},{0,0,0}]
Out[8]=
   {1}

In[9]:=
   extra[{1,0},{0,0},{0,0,0}]
Out[9]
   {1}
```

```
In[10]:=
   extra[{3,0},{2,3},{1,2,3}]
Out[10]=
             n    2      n           2 n
    (q - b q ) (q  - c q ) (q - c q   )
   {---------------------------------}
                          n        2
       (b - q) (c - q) q  (-c + q )

In[11]:=
   extra[{0,0},{0,-1},{0,0,0}]
Out[11]=
                  1 + 2 n
        1 - c q
   {-(------------------)}
        n            1 + n
      q  (-1 + c q     )
```

(*demonstration for $\chi_1 \begin{bmatrix} \alpha, \gamma; \varepsilon \\ b, c; n \end{bmatrix} \{\lambda, \mu, \nu\} \end{bmatrix}$ *)

```
In[12]:=
   extra[{0,1},{0,0},{0,0,0}]
Out[12]=
        -1 - n      1 + n
   {-(q        (1 - q    ))}

In[13]:=
   extra[{1,1},{0,0},{0,0,0}]
Out[13]=
   {0}

In[14]:=
   extra[{2,1},{0,0},{0,0,0}]
Out[14]=
     n      1 + n
   {q  (1 - q    )}

In[15]:=
   extra[{1,1},{1,-1},{0,1,0}]
Out[15]=
                       1 + n
      (b - c q) (1 - q    )
   {-----------------------}
                       1 + n
      (-b + q) (-1 + c q    )
```

```
In[16]:=
   extra[{1,1},{0,1},{0,0,1}]
Out[16]=
            1 + n
      1 - q
   {-(-----------)}
        n
      q  (-c + q)

In[17]:=
   extra[{2,1},{0,1},{0,0,1}]
Out[17]=
          n        1 + n
      c q  (1 - q     )
   {-(------------------)}
             -c + q

In[18]:=
   extra[{0,1},{0,-1},{0,0,0}]
Out[18]=
                 1 + n
      c (1 - q     )
   {-(--------------)}
                 1 + n
      -1 + c q

In[19]:=
   extra[{1,1},{0,-1},{0,0,0}]
Out[19]=
                 1 + n
      1 - q
   {-(-------------)}
                 1 + n
      -1 + c q

In[20]:=
   extra[{2,1},{1,1},{0,1,1}]
Out[20]=
                 n        1 + n
      (-q + b c q ) (1 - q     )
   {---------------------------}
         (-b + q) (-c + q)
```

```
In[21]:=
    extra[{0,1},{-1,-1},{-1,-1,-1}]
              1 + n              1 + n
       (-1 + q    ) (1 - b c q     )
    {-(-------------------------------)}
              1 + n             1 + n
       (1 - b q    ) (1 - c q     )

In[22]:=
    extra[{2,1},{0,2},{0,0,2}]
              n          n        1 + n
       (-q - c q ) (q - c q ) (1 - q     )
    {---------------------------------}
              n          2
          q  (-c + q) (-c + q )

In[23]:=
    extra[{0,1},{0,-2},{0,0,0}]
Out[23]=
              1 + n             1 + n
       (1 - q     ) (-1 - c q     )
    {---------------------------}
              n          2 + n
          q  (-1 + c q     )


In[24]:=
    extra[{3,1},{2,2},{0,2,2}]
Out[24]=
              n          n     2        n          1 + n        1 + n
       (q - b q ) (q - c q ) (-q  + b c q ) (-1 - q     ) (1 - q     )
    {-(-----------------------------------------------------------)}
              n                        2          2
          q  (-b + q) (-c + q) (-b + q ) (-c + q )

In[25]:=
    extra[{-1,1},{-2,-2},{-2,-2,-2}]
Out[25]:=
              1 + n          1 + n             2 + n
       (-1 - q    ) (-1 + q    ) (1 - b c q     )
    {---------------------------------------}
              n          2 + n          2 + n
          q  (1 - b q    ) (1 - c q     )
```

References

[1] W. N. Bailey, *Generalized Hypergeometric Series*, Cambridge University Press, Cambridge, 1935.

[2] W. N. Bailey, *A note on certain q-identities*, Quart. J. Math. Oxford 12 (1941), 173-175.

[3] W. N. Bailey, *On the analogue of Dixon's theorem for bilateral basic hypergeometric series*, Quart. J. Math. Oxford (2) 1 (1950), 318-320.

[4] D. M. Bressoud, *Almost poised basic hypergeometric series*, Proc. Indian Acad. Sci. (Math. Sci.) 97:1-3 (1987), 61-66.

[5] L. Carlitz, *Some formulas of F. H. Jackson*, Monatshhefte für Math.73 (1969), 193-198.

[6] L. Carlitz, *Some inverse relations*, Duke Math. J.40 (1973), 893-901.

[7] CHU Wenchang, *Natural q-analogues of the Gould-Hsu inversions*, Math. Appl.2:1 (1989), 47-52.

[8] CHU Wenchang, *Inversion techniques and combinatorial identities*, Bollettino U. M. I.(7) 7-B (1993), 737-760.

[9] CHU Wenchang, *Partial fractions and bilateral summations*, J. Math. Phys.35:4 (1994), 2036-2042; *Erratum*, ibid 36:9(1995), 5198-5199.

[10] G. Gasper, *Summation formulas for basic hypergeometric series*, SIAM J. Math. Anal.12 (1981), 196-200.

[11] G. Gasper & M. Rahman, *Basic Hypergeometric Series*, Vol.35 in Encyclopedia of Mathematics and Its Applications (edited by G. C. Rota), Cambridge University Press, Cambridge, 1990.

[12] R. L. Graham, D. E. Knuth & O. Patashnik, *Concrete Mathematics*, Addison-Wesley, Reading Mass, 1989.

[13] F. H. Jackson, *Summation of q-hypergeometric series*, Messenger of Math. 50 (1921), 101-112.

[14] F. H. Jackson, *Certain q-identities*, Quart. J. Math. Oxford 12 (1941), 167-172.

[15] V. K. Jain & A. Verma, *Certain summation formulae for q-series*, J. Indian Math. Soc.47 (1983), 71-85.

[16] C. M. Joshi & A. Verma, *Some remarks on summation of basic hypergeometric series*, Houston J. Math.5:2 (1979), 277-294.

[17] L. J. Slater, *Generalized Hypergeometric Functions*, Cambridge University Press, Cambridge, 1966.

Editorial Information

To be published in the *Memoirs*, a paper must be correct, new, nontrivial, and significant. Further, it must be well written and of interest to a substantial number of mathematicians. Piecemeal results, such as an inconclusive step toward an unproved major theorem or a minor variation on a known result, are in general not acceptable for publication. *Transactions* Editors shall solicit and encourage publication of worthy papers. Papers appearing in *Memoirs* are generally longer than those appearing in *Transactions* with which it shares an editorial committee.

As of June 30, 1998, the backlog for this journal was approximately 9 volumes. This estimate is the result of dividing the number of manuscripts for this journal in the Providence office that have not yet gone to the printer on the above date by the average number of monographs per volume over the previous twelve months, reduced by the number of issues published in four months (the time necessary for preparing an issue for the printer). (There are 6 volumes per year, each containing at least 4 numbers.)

A Copyright Transfer Agreement is required before a paper will be published in this journal. By submitting a paper to this journal, authors certify that the manuscript has not been submitted to nor is it under consideration for publication by another journal, conference proceedings, or similar publication.

Information for Authors and Editors

Memoirs are printed by photo-offset from camera copy fully prepared by the author. This means that the finished book will look exactly like the copy submitted.

The paper must contain a *descriptive title* and an *abstract* that summarizes the article in language suitable for workers in the general field (algebra, analysis, etc.). The *descriptive title* should be short, but informative; useless or vague phrases such as "some remarks about" or "concerning" should be avoided. The *abstract* should be at least one complete sentence, and at most 300 words. Included with the footnotes to the paper, there should be the 1991 *Mathematics Subject Classification* representing the primary and secondary subjects of the article. This may be followed by a list of *key words and phrases* describing the subject matter of the article and taken from it. A list of the numbers may be found in the annual index of *Mathematical Reviews*, published with the December issue starting in 1990, as well as from the electronic service e-MATH [**telnet e-MATH.ams.org** (or **telnet 130.44.1.100**). Login and password are **e-math**]. For journal abbreviations used in bibliographies, see the list of serials in the latest *Mathematical Reviews* annual index. When the manuscript is submitted, authors should supply the editor with electronic addresses if available. These will be printed after the postal address at the end of each article.

Electronically prepared papers. The AMS encourages submission of electronically prepared papers in $\mathcal{A}_{\mathcal{M}}\mathcal{S}$-TeX or $\mathcal{A}_{\mathcal{M}}\mathcal{S}$-LaTeX. The Society has prepared author packages for each AMS publication. Author packages include instructions for preparing electronic papers, the *AMS Author Handbook*, samples, and a style file that generates the particular design specifications of that publication series for both $\mathcal{A}_{\mathcal{M}}\mathcal{S}$-TeX and $\mathcal{A}_{\mathcal{M}}\mathcal{S}$-LaTeX.

Authors with FTP access may retrieve an author package from the Society's Internet node `e-MATH.ams.org` (130.44.1.100). For those without FTP access, the author package can be obtained free of charge by sending e-mail to `pub@ams.org` (Internet) or from the Publication Division, American Mathematical Society, P.O. Box 6248, Providence, RI 02940-6248. When requesting an author package, please specify $\mathcal{AMS}$-TEX or $\mathcal{AMS}$-LATEX, Macintosh or IBM (3.5) format, and the publication in which your paper will appear. Please be sure to include your complete mailing address.

Submission of electronic files. At the time of submission, the source file(s) should be sent to the Providence office (this includes any TEX source file, any graphics files, and the DVI or PostScript file).

Before sending the source file, be sure you have proofread your paper carefully. The files you send must be the EXACT files used to generate the proof copy that was accepted for publication. For all publications, authors are required to send a printed copy of their paper, which exactly matches the copy approved for publication, along with any graphics that will appear in the paper.

TEX files may be submitted by email, FTP, or on diskette. The DVI file(s) and PostScript files should be submitted only by FTP or on diskette unless they are encoded properly to submit through e-mail. (DVI files are binary and PostScript files tend to be very large.)

Files sent by electronic mail should be addressed to the Internet address `pub-submit@ams.org`. The subject line of the message should include the publication code to identify it as a Memoir. TEX source files, DVI files, and PostScript files can be transferred over the Internet by FTP to the Internet node `e-math.ams.org` (130.44.1.100).

Electronic graphics. Figures may be submitted to the AMS in an electronic format. The AMS recommends that graphics created electronically be saved in Encapsulated PostScript (EPS) format. This includes graphics originated via a graphics application as well as scanned photographs or other computer-generated images.

If the graphics package used does not support EPS output, the graphics file should be saved in one of the standard graphics formats—such as TIFF, PICT, GIF, etc.—rather than in an application-dependent format. Graphics files submitted in an application-dependent format are not likely to be used. No matter what method was used to produce the graphic, it is necessary to provide a paper copy to the AMS.

Authors using graphics packages for the creation of electronic art should also avoid the use of any lines thinner than 0.5 points in width. Many graphics packages allow the user to specify a "hairline" for a very thin line. Hairlines often look acceptable when proofed on a typical laser printer. However, when produced on a high-resolution laser imagesetter, hairlines become nearly invisible and will be lost entirely in the final printing process.

Screens should be set to values between 15% and 85%. Screens which fall outside of this range are too light or too dark to print correctly.

Any inquiries concerning a paper that has been accepted for publication should be sent directly to the Editorial Department, American Mathematical Society, P. O. Box 6248, Providence, RI 02940-6248.

Editors

This journal is designed particularly for long research papers (and groups of cognate papers) in pure and applied mathematics. Papers intended for publication in the *Memoirs* should be addressed to one of the following editors:

Ordinary differential equations, partial differential equations, and applied mathematics to JOHN MALLET-PARET, Division of Applied Mathematics, Brown University, Providence, RI 02912-9000; electronic mail: `jmp@cfm.brown.edu`.

Harmonic analysis, representation theory, and Lie theory to ROBERT J. STANTON, Department of Mathematics, The Ohio State University, 231 West 18th Avenue, Columbus, OH 43210-1174; electronic mail: `stanton@math.ohio-state.edu`.

Ergodic theory and dynamical systems to ROBERT F. WILLIAMS, Department of Mathematics, University of Texas at Austin, Austin, TX 78712-1082; e-mail: `bob@math.utexas.edu`

Real and harmonic analysis and geometric partial differential equations to WILLIAM BECKNER, Department of Mathematics, University of Texas at Austin, Austin, TX 78712-1082; e-mail: `beckner@math.utexas.edu`.

Algebra to CHARLES CURTIS, Department of Mathematics, University of Oregon, Eugene, OR 97403-1222 e-mail: `cwc@darkwing.uoregon.edu`

Algebraic topology and cohomology of groups to STEWART PRIDDY, Department of Mathematics, Northwestern University, 2033 Sheridan Road, Evanston, IL 60208-2730; e-mail: `s_priddy@math.nwu.edu`.

Differential geometry and global analysis to CHUU-LIAN TERNG, Department of Mathematics, Northeastern University, Huntington Avenue, Boston, MA 02115-5096; e-mail: `terng@neu.edu`.

Probability and statistics to RODRIGO BAÑUELOS, Department of Mathematics, Purdue University, West Lafayette, IN 47907-1968; e-mail: `banuelos@math.purdue.edu`.

Combinatorics and Lie theory to PHILIP J. HANLON, Department of Mathematics, University of Michigan, Ann Arbor, MI 48109-1003; e-mail: `hanlon@math.lsa.umich.edu`.

Logic to THEODORE SLAMAN, Department of Mathematics, University of California at Berkeley, Berkeley, CA 94720-3840; e-mail: `slaman@math.berkeley.edu`.

Number theory and arithmetic algebraic geometry to ALICE SILVERBERG, c/o Department of Mathematics, Evans Hall, University of California at Berkeley, Berkeley, CA 94720; e-mail: `silver@math.ohio-state.edu`.

Complex analysis and complex geometry to DANIEL M. BURNS, Department of Mathematics, University of Michigan, Ann Arbor, MI 48109-1003; e-mail: `dburns@math.lsa.umich.edu`.

Algebraic geometry and commutative algebra to LAWRENCE EIN, Department of Mathematics, University of Illinois, 851 S. Morgan (M/C 249), Chicago, IL 60607-7045; e-mail: `ein@uic.edu`.

Geometric topology, knot theory, hyperbolic geometry, and general topoogy to JOHN LUECKE, Department of Mathematics, University of Texas at Austin, Austin, TX 78712-1082; e-mail: `luecke@math.utexas.edu`.

Partial differential equations and applied mathematics to BARBARA LEE KEYFITZ, Department of Mathematics, University of Houston, 4800 Calhoun, Houston, TX 77204-3476; e-mail: `keyfitz@uh.edu`

Operator algebras and functional analysis to BRUCE E. BLACKADAR, Department of Mathematics, University of Nevada, Reno, NV 89557; e-mail: `bruceb@math.unr.edu`

All other communications to the editors should be addressed to the Managing Editor, PETER SHALEN, Department of Mathematics, University of Illinois, 851 S. Morgan (M/C 249), Chicago, IL 60607-7045; e-mail: `shalen@math.uic.edu`.